Energy Efficiency Guide for Existing Commercial Buildings: The Business Case for Building Owners and Managers

---

Lead Authors
Dennis R. Landsberg, L&S Energy Services, Inc.
Mychele R. Lord, LORD Green Real Estate Strategies, Inc.

Contributing Authors
Steven Carlson, CDH Energy Corp.
Fredric S. Goldner, Energy Management & Research Associates

Developers
American Society of Heating, Refrigerating
and Air-Conditioning Engineers, Inc.
The American Institute of Architects
Illuminating Engineering Society of North America
U.S. Green Building Council

Contributors
Building Owners and Managers Association International
U.S. General Services Administration

ISBN 978-1-933742-63-2

1791 Tullie Circle, NE
Atlanta, GA 30329
www.ashrae.org

Printed in the United States of America

Cover design by Emily Luce, Designer.
Cover photographs courtesy of H.E. Burroughs

Library of Congress Cataloging-in-Publication Data

Landsberg, Dennis R., 1948-

Energy efficiency guide for existing commercial buildings : the business case for building owners and managers / lead authors, Dennis R. Landsberg, Mychele R. Lord ; contributing authors, Steven Carlseo, Fredric Goldner ; developers, American Society of Heating, Refrigerating, and Air-Conditioning Engineers, Inc. ... [et al.] ; contributors, Building Owners and Managers Association International [and] U.S. General Services Administration.

p. cm.

Includes bibliographical references.

Summary: "This guide explains why building owners and managers should be concerned about energy efficiency. The first in a series, the goal of this resource is to help this audience evaluate current building operations and conduct the economic analysis necessary to make decisions to improve energy efficiency"--Provided by publisher.

ISBN 978-1-933742-63-2 (softcover : alk. paper) 1. Commercial buildings--Energy conservation. 2. Commercial buildings--Energy consumption. 3. Commercial buildings--Cost of operation. I. Lord, Mychele R. II. Carlson, S. W. (Steven W.) III. American Society of Heating, Refrigerating and Air-Conditioning Engineers. IV. Title.

TJ163.5.B84L355 2009

658.2'6--dc22

2009030155

Contents

Acknowledgments

The *Energy Efficiency Guide for Existing Commercial Buildings: The Business Case for Building Owners and Managers* is the first in a series of three Guides that address the existing commercial building stock in the United States.

The main contributors to the Guide were the primary authors, Dennis R. Landsberg and Mychele R. Lord, and the contributing authors, Steven Carlson and Fredric Goldner. These authors put in many long hours and worked collaboratively to produce this Guide.

The Advanced Energy Design Guide (AEDG) steering committee provided direction and guidance to complete this manuscript and produced an invaluable scoping document to begin the creative process. Members on the project monitoring committee came from the AEDG steering committee partner organizations as well as the Building Owners and Managers Association International and the U.S. General Services Administration. These members served not only on the project monitoring committee but also as liaisons to their respective organizations.

The chairman would like to personally thank the authors and all of the members of the project monitoring committee for their diligence, creativity, and persistence.

In addition to the authors and the members on the committee, there were a number of other individuals who contributed to the success of this Guide. The specific individuals and their contributions were: Bruce Hunn and Lilas Pratt of ASHRAE for their assistance, organizational skills, and dedication to the project; Amelia Sanders of ASHRAE Special Publications for editing and laying out this book; and Cindy Michaels of ASHRAE Special Publications for proofreading. This Guide could not have been developed without all of their contributions.

I am very proud of the Guide that has been developed and amazed at what has been accomplished in such a short time period. The authors and each of the project monitoring committee members should be proud of their individual contributions to this most worthwhile document.

George Jackins
SP-118 Chair
June 2009

Abbreviations & Acronyms

AEDG	=	*Advanced Energy Design Guide*
AIA	=	American Institute of Architects
ANSI	=	American National Standards Institute
ASHRAE	=	American Society of Heating, Refrigerating and Air-Conditioning Engineers
BEEP	=	Building Energy and Efficiency Program (developed by BOMA International)
BOMA	=	Building Owners and Managers Association
Btu	=	British thermal unit
CBEC	=	Commercial Buildings Energy Consumption Survey
CDD	=	cooling degree-day
CFL	=	compact fluorescent lamp
CO_2	=	carbon dioxide
DOE	=	U.S. Department of Energy
EEG-EB	=	*Energy Efficiency Guide for Existing Commercial Buildings*
EEM	=	energy efficiency measure
EIA	=	Energy Information Administration
EMS	=	energy management systems
EPA	=	U.S. Environmental Protection Agency
EPEAT	=	Electronic Product Environmental Assessment Tool
ESCO	=	energy service company
ESPC	=	energy saving performance contracting
EUI	=	energy utilization index, $Btu/ft^2 \cdot yr$
FEMP	=	Federal Energy Management Program (developed by the National Institute of Standards and Technology)
GSA	=	U.S. General Services Administration
HDD	=	heating degree-day
HVAC	=	heating, ventilating, and air-conditioning
IEA	=	International Energy Agency
IES	=	Illuminating Engineering Society of North America
IFMA	=	International Facility Management Association

IGA	=	investment-grade energy audit
IRR	=	internal rate of return
IPCC	=	Intergovernmental Panel on Climate Change
kW	=	kilowatt
kWh	=	kilowatt hour
LBNL	=	Lawrence Berkeley National Laboratory
LCC	=	life-cycle costing
LED	=	light-emitting diode
LEED	=	Leadership in Energy and Environmental Design
M&V	=	measurement and verification
NAESCO	=	National Association of Energy Service Companies
NIST	=	National Institute of Standards and Technology
NPV	=	net present value
NYSIO	=	New York Independent System Operator
O&M	=	operations and maintenance
PC	=	project committee
PV	=	photovoltaic
SC	=	steering committee
SEER	=	seasonal energy efficiency ratio
SP	=	special project
TC	=	technical committee
USGBC	=	U.S. Green Building Council
W	=	watts

Preface

There is a significant long-term asset value that can be placed on buildings that have been designed and can operate as energy efficient, green, and sustainable. The surging interest in environmentally friendly building programs represents a tremendous opportunity for building owners and real estate developers—with the great potential for a positive return on the investment.

This is the first in a series of Energy Efficiency Guides (EEGs) for existing buildings and is aimed at providing the business case for energy efficiency to building owners and managers. The Guide provides the rationale for making economic decisions related to improving and sustaining energy efficiency in existing buildings. The goal is to enable building owners to undertake the processes for evaluating current operations and do the economic analysis of options for improvement. The Guide demonstrates ways to benchmark performance against comparable buildings; illustrates ways under which energy use and cost can be reduced by at least 30%; and details how this can be done, all while minimizing capital investment and maximizing return on investment. It further identifies additional measures that can achieve even greater energy efficiencies and cost savings in economical ways.

This Guide will be followed by two other EEGs. The first of the two will be aimed at providing technical guidance in undertaking existing building renovation programs, and the second will provide operational and maintenance guidance to sustain the energy efficiency and performance of buildings.

American Society of Heating, Refrigerating and Air-Conditioning Engineers, Inc. (ASHRAE), in collaboration with the American Institute of Architects (AIA), the Illuminating Engineering Society of North America (IES), and the U.S. Green Building Council (USGBC), with support from the U.S. Department of Energy (DOE), has published a series of highly successful Advanced Energy Design Guides (AEDGs). These guides provide prescriptive guidance for new building designs to achieve energy savings 30% beyond those described in *ANSI/ASHRAE/IESNA Standard 90.1, Energy Standard for Buildings Except Low-Rise Residential Buildings*

(ASHRAE 2007d). The guides include the *Advanced Energy Design Guide for Small Office Buildings* (ASHRAE 2005a), *Advanced Energy Design Guide for Small Retail Buildings* (ASHRAE 2006a), *Advanced Energy Design Guide for K–12 School Buildings* (ASHRAE 2007b), *Advanced Energy Design Guide for Small Warehouses and Self-Storage Buildings* (ASHRAE 2008), and the *Advanced Energy Design Guide for Highway Lodging* (ASHRAE 2009). The final guide in the 30% series, the *Advanced Energy Design Guide for Small Hospitals and Healthcare Buildings,* has a planned publication of November 2009.

This AEDG series is designed to lead the way in developing energy-efficient building design tools that are practical (using off-the-shelf technologies) while at the same time being economically viable. They dovetail well with other guidance provided by ASHRAE standards and guidelines, yet, realistically, all of these efforts primarily target new construction or, to a limited extent, portions of existing buildings undergoing major renovation.

Unlike the AEDGs, which were written by teams of volunteers from the collaborating organizations, the EEGs are being written by contractor(s), under the guidance of an EEG-EB Project Committee. In addition to the collaborating partner organizations, the Building Owners and Managers Association (BOMA) International and the U.S. General Services Administration (GSA) were also involved in the development of this Guide and were represented on the project committee.

When looking at our overall actions to reduce energy use, it is critical to recognize that new construction only represents about 2% of building programs. Conversely, approximately 86% of U.S. annual building construction expenditures relate to the renovation of existing buildings rather than new construction. Even more importantly, analysis of many of those new buildings indicates that their performance significantly deteriorates in the first three years of operation (some say by as much as 30%)—even those designed as high-energy-efficient green buildings. Research has also shown the potential for a 10% to 40% reduction in energy use simply by changing operational strategies.

As we look toward a more sustainable future, it should also be recognized that 70% to 85% of buildings that will exist in the year 2030 exist today. At the same time, it is estimated that over the next 30 years, about 150 billion ft^2 of existing buildings (roughly half of the entire building stock in the United States) will need to be renovated.

Clearly, the greatest opportunity for overall reduction in U.S. primary energy use lies within the existing building stock. That stock also represents a significant potential for real estate owners and developers to not only demonstrate sustainability initiatives but also to realize a great return on the investment. While ASHRAE has a significant number of initiatives already under way, it is now developing an overall, cohesive program aimed specifically at promoting energy conservation and efficiency in existing buildings and providing guidance to the real estate industry.

Gordon Holness, 2009–2010 ASHRAE President
May 2009

Introduction 1

This Guide has been developed for building owners, managers, operators, tenants, investors, and lenders to provide impartial information, tools, and resources to equip decision makers to act now to make their buildings more energy efficient. It is both economically responsible and ethically justifiable to actively manage the energy demand and carbon impact of all building operations. The actions recommended in this Guide will enable building owners to increase the operational performance of their buildings.

Increasing the energy efficiency of commercial and residential buildings is the nation's foremost low-cost climate change control strategy in a world that is heavily dependent upon carbon-producing energy generation. But, the United States faces serious challenges with its energy system. For example, demand for energy continues to rise despite volatile prices, mounting concern over energy security and independence, the need for significant capital investment in new energy infrastructure, air pollution, and global climate change.

Climate change policy, in the form of local and state green building and energy efficiency codes, land use policy, and carbon regulation, has emerged in recent years. The urgency to reduce greenhouse gas emissions is already impacting the building sector through tenant space requirements, local building codes, and the surge in green buildings. Plus, proposed carbon legislation at the federal level may impact building operating costs if enacted into law.

Opportunities for improved energy efficiency are underrealized in existing buildings, particularly in the investment property sector. This is because the developer that pays the first cost or the landlord that makes a capital investment is often not the direct beneficiary of the resulting reduced operating costs. However, this is changing as more and more tenants are seeking energy-efficient, green buildings and incorporating a building's energy performance into their site selection criteria.

Owners who authorize energy audits and energy-focused retro-commissioning, who seize opportunities to improve energy efficiency through operational adjustments, are winning the battle against rising prices. Plus, new technologies that make equipment more efficient, coupled with rising energy costs, significantly improve the financial returns on investment in new equipment. And for planned replacements, tax credits are available for new energy-efficient technologies and on-site power generation. The bottom line: energy-efficient, green buildings aren't just good for the environment; they are good for business.

When updating and relocating their Atlanta offices, the Swedish construction giant Skanska surmounted the challenges of incorporating sustainable principles into its leased space by working with the landlord during the early stages of the project, starting with a comprehensive plan developed through a partnership with Jova/Daniels/Busby, the architects for the project.

(Photograph courtesy of Jova/Daniels/Busby. Photographer: Robert Thien, Inc.)

ESCALATING ENERGY PRICES

A recent report from the International Energy Agency (IEA) indicates the world's primary energy needs are expected to grow by 55% between 2005 and 2030 (IEA 2007). So, of course this is going to impact the cost of energy. For example, a 2008 International Facility Management Association (IFMA) report, which gathers data from more than 1000 survey responses, reported a 19% increase in utility costs since 2006 (IFMA 2008). Also, a review of operating expense data among BOMA International members since 2004 revealed that increases in the average electricity cost per square foot for office buildings is outpacing total expense growth. Further, utility expenditures are the largest expense after taxes—that is, the largest, most controllable operating expense (a fifth of total costs and about a third of total variable costs) (BOMA 2004–2007). (See Figure 1-1.)

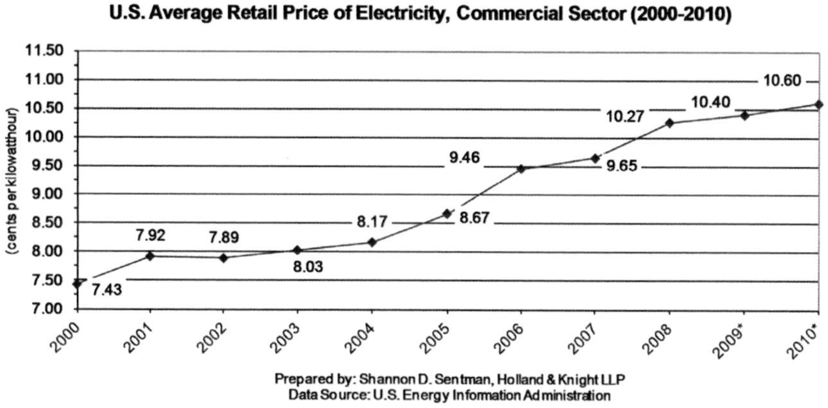

Figure 1-1. The increase in average retail price for commercial sector end users from 2000 through 2007, with projections for 2008, 2009, and 2010 (EIA 2007, 2009b).

Once existing building energy efficiency is improved, owners will enjoy positive changes when it comes to their bottom line, environment, and public/tenant perceptions.

ENERGY EFFICIENCY—A CLIMATE CHANGE STRATEGY

The United States represents less than 5% of the world's population but contributes approximately 25% of the world's greenhouse gas emissions. According to the U.S. Department of Energy (DOE), more than 49% of electricity is generated from coal, with buildings accounting for more than 70% of the electricity load in the United States. Buildings as end users are significant contributors to total carbon emissions. Buildings account for more than 39% of the country's greenhouse gas emissions (18% for commercial and 21% for residential) compared with transportation (29%) and industry (32%) (see Figure 1-2). Commercial buildings are the only sector that has outpaced gross domestic product in emissions growth (EIA 2008).

Recent studies by McKinsey & Company (2007), the Intergovernmental Panel on Climate Change (IPCC) (2007), and a review of 80 studies on 5 continents (UNEP FI 2008) found the greatest opportunity for cost-effective carbon dioxide (CO_2) reduction in developed countries is to increase the energy efficiency of buildings. With existing buildings outnumbering new buildings by more than 50 to 1, knowing and improving the energy efficiency of our existing building stock is paramount to achieving CO_2 reduction.

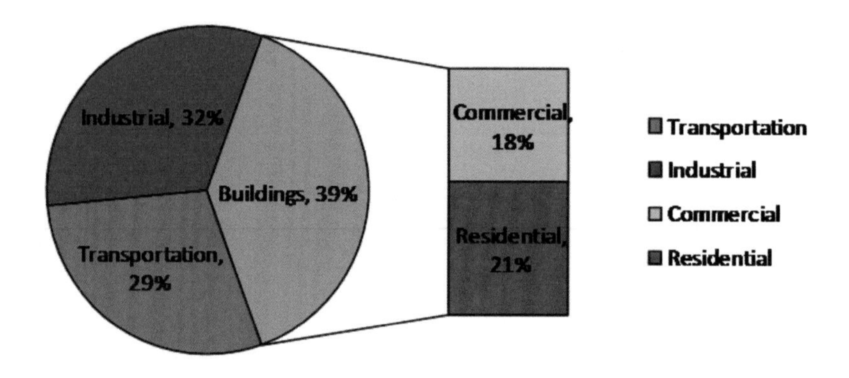

Figure 1-2. Greenhouse gas emissions.

CORPORATE RESPONSIBILITY AND SUSTAINABILITY REPORTING

Measuring and reporting the environmental impact of real estate decisions falls within what is often called *corporate responsibility and sustainability* (i.e., acting in a fiscally, socially, and environmentally responsible manner) (REALpac 2007). Green has gone mainstream and, as a result, increasingly corporations are engaging voluntarily in corporate responsibility and sustainability reporting. An important part of reporting is tracking carbon emissions using methods such as the Greenhouse Gas Protocol (WRI and WBCSD 1998). Reporting carbon emissions involves establishing a baseline and setting reduction goals. Factors for calculating the emissions of an organization include, at minimum, emissions resulting from operations, employee commutes, and real estate.

CARBON REDUCTION INITIATIVES/REQUIREMENTS

As of August 2, 2009, 965 mayors have signed the U.S. Conference of Mayors Climate Protection Agreement (2009). They recognize that air quality is critical to the continued economic development of their cities. Building energy efficiency improves air quality and reduces the need for additional energy production and distribution capacity.

The American Institute of Architects (AIA) and the Conference of Mayors have adopted Architect 2030's 2030 Challenge (2009), which calls for an immediate 50% reduction in fossil fuel consumption for new buildings compared with the regional average, while moving toward net zero energy buildings by 2030. As a result, local building energy codes are becoming more and more stringent in order to meet the 2030 Challenge. The Energy Independence and Security Act (2007) signed by President Bush in December 2007 establishes a Net Zero Energy Commercial Buildings Initiative and sets goals for newly constructed commercial buildings to be zero net energy by 2030 and the same for 50% of the commercial building stock by 2040 and all commercial buildings by 2050.

California's Assembly Bill 1103 (2007) requires, as of January 1, 2010, that a nonresidential building owner or operator disclose ENERGY STAR Portfolio Manager energy performance ratings (EPA 2008b) and data to a prospective buyer, lessee, or lender as part of a whole-building transaction—similar to the requirements in the European Union. The District of Columbia's Clean & Affordable Energy Act of 2008 (2008) requires commercial buildings to be rated using ENERGY STAR Portfolio Manager by 2010 and for the information to be made available to the public by the District Department of the Environment.

STRATEGIES FOR IMPROVING ENERGY EFFICIENCY

The American Society of Heating, Refrigerating and Air-Conditioning Engineers, Inc. (ASHRAE) has developed guideline documents to assist building owners and operators in identifying and developing energy-efficiency improvements.

The Building Owners and Managers Association (BOMA) International and ENERGY STAR have identified significant energy savings that can be obtained through EEMs. Specifically, the BOMA International Building Energy Efficiency Program (BEEP) reports the following energy savings potential (BOMA 2006a):

- 7% to 28% energy savings can be achieved by implementing no- and low-cost EEMs through changes in O&M.
- An additional 3.5% to 15.2% energy savings can be recognized through changes in occupant behavior such as instituting an energy awareness program, turning off equipment, purchasing ENERGY STAR-labeled equipment, installing power management software, and using task/ambient lighting instead of general area lighting.
- Lighting is another area where building operators can achieve dramatic financial returns with low capital investment and use off-the-shelf, proven technologies. Lighting accounts for approximately 29% of energy used in office buildings. The installation of current, commercially available technologies often has less than a one-year simple payback. Cumulative energy savings from lighting could range from 9.4% to 25%.
- Control devices can be calibrated and monitored to more effectively reduce energy consumption from 7.3% to 22.9%.
- Building owners and managers should also examine potential equipment changes for heating, ventilating, and air-conditioning (HVAC) systems and controls. The whole-building energy savings potential for equipment changes ranges from 3.5% to 15.9% .

There are many energy-efficiency methods that operators of public- and private-sector buildings can use to improve the performance of the buildings, reduce energy consumption, and save money.

Further, if capital-intensive upgrades are scheduled for a building, owners shouldn't replace like with like. Instead, they should use NPV or LCC techniques to evaluate new technologies that bring greater energy efficiency.

The Steps to Take

Take the following steps to identify and improve the energy performance of the buildings you own or operate.

- Measure the energy performance and carbon emissions of your buildings.
- Evaluate building energy performance, and set goals that are in line with asset or company strategy.
- Implement no- and low-cost operational adjustments to improve operational efficiency.
- Conduct an ASHRAE Level I Walk-Through Energy Audit and/or hire an independent third-party to conduct an ASHRAE Level II Energy Audit or energy-focused retrocommissioning of the building.
- Evaluate, develop a plan for, and implement EEMs identified in the audits or retrocommissioning.
- Take an integrated, forward-thinking approach to planned replacements—staging replacements and evaluating new energy-efficient technologies, on-site power generation, and available tax credits.
- Incorporate energy management into standard policies and procedures—procurement, O&M, preventative maintenance—and ensure ongoing performance tracking.

HOW TO USE THIS GUIDE

Energy Efficiency Guide for Existing Commercial Buildings: The Business Case for Building Owners and Managers (EEG-EB) provides a road map to increased energy efficiency in existing buildings. Application of the EEG-EB should, at minimum, result in reduced energy consumption of 30% or more for the average building.

This Guide has been developed under the supervision of a diverse committee of energy professionals drawn from the following organizations:

- ASHRAE
- BOMA International
- Illuminating Engineering Society of North America
- U.S. General Services Administration
- U.S. Green Building Council

These professionals came together under the leadership of the Advanced Energy Design Guide Steering Committee to develop this Guide, which is meant to provide building owners and operators with an objective resource for continually improving the energy efficiency of existing buildings. The following are summaries of what will be covered in each chapter.

Chapter 2—Measuring Energy Efficiency and Setting Goals

Managing a building's energy efficiency may soon be an integral component of managing its operational and financial performance. The first step is to characterize the energy efficiency of the building. Energy efficiency for buildings is measured in terms of an energy utilization index (EUI). An EUI is made relevant by comparing the EUI with known building population EUI distributions or through performance rating tools. Building population EUI distributions can come from either inside or outside of an organization (i.e., internal or external reference). Weather normalization is also usually important when comparing EUIs over different years, as energy use can change up to 15% with severe vs. mild summers and winters. Energy performance rating tools typically normalize for weather and other parameters such as occupancy density and operating hours. If one is only comparing EUIs to population distributions, additional weather normalization may be needed. ENERGY STAR Portfolio Manager (EPA 2008b) normalizes a building's EUI for several factors, including weather, and then reports energy performance on a scale of 1–100. Once a building's energy performance is understood, the next step is to develop meaningful and attainable performance goals. Goals could include trying to reduce energy consumption by 10% over 12 months, reducing a building's carbon emissions, or achieving the ENERGY STAR label.

Chapter 3—Improving Energy Performance

Once a building's energy performance is established and performance goals are set, the next step is to identify how to improve a building's energy efficiency. The primary end uses of energy in a building are lighting, plug loads (i.e., appliances that are plugged into wall outlets), heating, ventilating and air conditioning, and domestic water heating. Building operational and maintenance practices and occupant behaviors should be reviewed to identify the low-hanging fruit—energy-conserving actions that can be implemented at little or no cost. Energy audits and energy-focused retrocommissioning identify energy efficiency measures (EEMs). When making capital upgrades and repairs, taking an integrated approach to achieve optimum, cost-effective results cannot be overemphasized.

Chapter 4—Making the Business Case

Once no- or low-cost operational adjustments, maintenance practices, and occupant behavioral changes are implemented or completed, the next step is to evaluate the risk and cost/benefit of capital-intensive EEMs identified by energy audits or energy-focused retrocommissioning. Establishing the business case for these capital expenditures involves risk and financial analysis. Financial analysis can no longer be limited to simple payback or internal rate of return but must be expanded to include net present value (NPV) and life-cycle costing (LCC) capital budgeting techniques. LCC can also be used to evaluate alternative strategies such as: Determining whether an alternative that costs more but has lower energy and/or maintenance costs over time fits within building owner budgets and financial performance criteria.

Chapter 5—Developing an Effective O&M Program

Operational EEMs are the no- and low-cost measures for improving energy efficiency by 5% to 20% for most buildings. Operational EEMs are the operations in operations and maintenance (O&M). To capture this low-hanging fruit requires elevating the importance of energy efficiency within the organization and providing building operators with the necessary tools and training. Ongoing performance tracking provides the only assurance that capital investments are paying off and that optimal energy performance, the maximum life of building systems, and indoor environmental quality are continually achieved.

Measuring Energy Efficiency and Setting Goals 2

INTRODUCTION

Whether a person is being mindful of her budget or her carbon footprint, making an informed decision before purchasing a vehicle requires knowing its fuel efficiency or miles per gallon. The same is true of a building, which is why its energy utilization index (EUI) is used to evaluate its energy efficiency. In underwriting investments, assumptions are made about cash flow and risk is assessed. For the remainder of this Guide, the energy metric dialed into all building operations decisions will be the EUI. Knowing and reporting the EUIs of buildings via the ENERGY STAR Portfolio Manager (EPA 2008b) has been legislated in the District of Columbia and California and is a matter of practice for companies such as JCPenney; Marriott International, Inc.; Principal Global Investors; TIAA-CREF; and USAA Real Estate Company. (ENERGY

The Durst organization has been an ENERGY STAR partner since 2002 and regularly tracks performance through EPA's ENERGY STAR Portfolio Manager. Its Condé Nast Building at 4 Times Square was the nation's first green high-rise.
(Photograph courtesy of *High Performing Buildings* [Hinge and Winston 2008].)

STAR maintains a state and local legislation fact sheet located on their ENERGY STAR for Government Web page at www.energystar.gov/government.)

This chapter provides building owners and operators with the knowledge to measure and evaluate the energy efficiency of a building and to set target energy use or reduction goals. This chapter describes the metrics used in defining a building's energy performance utilizing an EUI and how to assess and compare this information to that of similar buildings or to other buildings that are owned or managed by the user. EUI comparison tools such as the Commercial Buildings Energy Consumption Survey (CBECS) tables (EIA 2003) and the Environmental Protection Agency (EPA) ENERGY STAR Portfolio Manager (2008b) are covered in length. Integral to calculating a building's carbon emissions, site and source energy are also discussed.

MEASURING ENERGY PERFORMANCE WITH AN ENERGY UTILIZATION INDEX

The EUI is a ratio of the energy use of a building to a normalizing value such as floor area. For most buildings, this value is very useful, as it enables owners and operators to compare buildings to one another. Other forms of EUI ratios include energy use per guest room (for hotels/motels) or energy use per item produced (for manufacturing facilities). Throughout this Guide, the EUI will be expressed in its most common form for buildings—$Btu/ft^2 \cdot yr$.

The first step in understanding the relative energy use of a building is to determine its EUI. Determining the EUI of a building tells the owner/operator how efficient or inefficient a building is relative to its peer group. Once this is done, the building's EUI can be used to compare it to a portfolio of similar buildings in a region or around the nation. The EPA's ENERGY STAR Portfolio Manager (2008b) does just that by taking a property's EUI and providing a scale of comparison—1 to 100—by building type.

To calculate the EUI for a building, the first step is to find the sum of all of the energy used by the building in a given year. Appropriately placed monitors and data loggers can be used to measure the energy use in a building (see Figure 2-1). Then, all energy types are converted to British thermal units (Btus), and the EUI is calculated using the sum of all energy sources. For example, if electricity is the only fuel source used in the building, kilowatt hours (kWh) can be summed to establish total energy used by the building. However, most buildings use two or more types of fuels. To track total energy use in a building, all of the building's various energy types need to be converted. Put simply, to compute the EUI for a building, add up the annual consumption of each fuel type, convert the energy use to Btus, add the fuel sources, and divide by the gross floor area of the conditioned building.

Determining the Gross Floor Area

While the conditioned floor area is based on the gross floor area of spaces within a facility that are actually heated and cooled by facility heating, ventilating, and air-conditioning (HVAC) systems, the gross floor area includes additional spaces that are not typically air conditioned. The gross area is measured based on the outside of the exterior walls or the midpoint of the wall separating buildings. (This differs from the leasable space, which is measured based on the midpoint of the walls.) Areas such as basements, penthouses, storerooms, mechanical rooms, and mezzanines should be included in the EUI calculation. However, parking garages, which are typically lighted and ventilated but not heated, should not be included. Areas provided with only lighting, or provided with only partial service such as freeze-prevention heating, should also not be included in the gross floor area. The energy use of these areas should be removed from the EUI calculation by submetering or estimation. Examples include covered walkways, porches, exterior terraces, and chimneys.

The Basic Steps to Obtaining an EUI

For conversion factors for other fuel types and additional supporting information, refer to *ANSI/ASHRAE Standard 105-2007, Standard Methods of Measuring,*

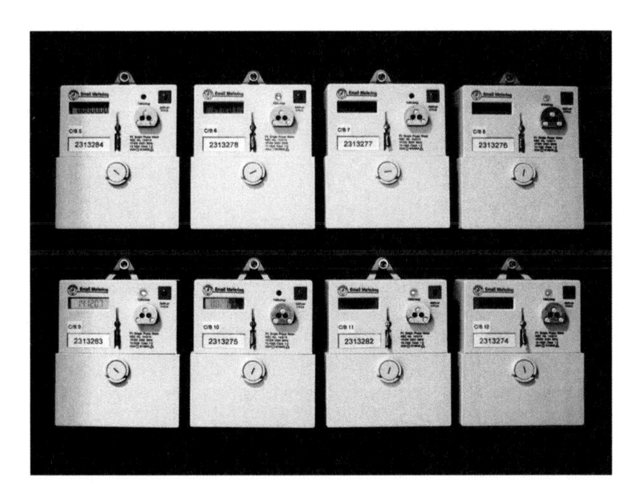

Figure 2-1. Monitors and data loggers installed throughout a building provide energy use data and can help reduce consumption.
(Photograph courtesy of *High Performing Buildings* [West 2009].)

Expressing, and Comparing Building Energy Performance (ASHRAE 2007e). The basic steps required to compute an EUI for a building are the following:

- Collect a minimum of 15 months' worth of historical energy consumption, demand, and cost data. This will ensure that a full year of representative data is utilized and anomalies are removed. A 24-month or more period is recommended to establish a year-to-year comparison (baseline reduction) in energy consumption, particularly if using tools such as the ENERGY STAR Portfolio Manager (EPA 2008b) to measure carbon equivalent reductions.
- Select a full-year period or close to a 365-day period from the data set (try to avoid including anomalies on either end).
- Calculate the Btus consumed for each energy source.
- Add all of the Btus together, arriving at the total annual Btus for the building.
- Calculate the EUI per square foot by dividing the total Btus by the gross floor area of the building.

Calculating the EUI is an ongoing process. The first full year for which the EUI is determined is often referred to as the *baseline* or *benchmark*. Loading energy data and calculating an EUI each month should become a disciplined practice and part of an organization's policies.

Table 2-1 demonstrates the calculation of the EUI value for a small building with three fuel types. Assuming a 20,000 ft^2 building uses 100,000 kWh of electricity, 2000 therms (a therm is 100,000 Btus) of natural gas, and 10,000 gal of #2 heating oil, the total annual building EUI would be 1,931,200,000 Btus.

GIVING MEANING TO AN EUI

To determine the energy performance of a building, the building's EUI is examined on a scale relative to comparable buildings. A number of factors impact

Table 2-1. Calculation of Building EUI

Fuel	**Usage**	**Conversion Factor (CF)**	**Usage × CF = Btus**
Electricity	100,000 kWh	3412 Btu/kWh	341,200,000
Natural Gas	2000 therms	100,000 Btu/therm	200,000,000
#2 Heating Oil	10,000 gal.	139,000 Btu/gal	1,390,000,000
Total Annual Building Energy Usage in Btus			1,931,200,000
Total Building Gross Square Footage			20,000
Total EUI			**96,560 Btu/ft²·yr**

energy use, such as building type, number of occupants, hours of use, and climate zone (i.e., California and North Dakota have widely differing climates). The comparison of a shopping mall to an office building is of little use. Similarly, the comparison of office buildings with widely different operating hours will likely be skewed and show more energy use in the building with longer operating hours. Therefore, not only does a data source for comparing buildings need to be defined, but variables must be isolated to obtain an accurate measure of energy performance relative to other buildings. These variables include type of building, weather, use, plug load, process load, operating hours, and number of occupants. If comparing buildings within a portfolio, inconsistencies across the portfolio have to be normalized to make a fair comparison.

If EUIs are calculated for multiple years, variations that arise from year to year will become apparent. There are two reasons for this. First, changes may have occurred in building operations, such as different operating hours, a change in tenancy, occupant use, manual override of controls by building operators, remodeling or equipment replacement, or deteriorating equipment and faulty controls. Second, the weather varies from year to year, and this causes a change in the energy required to heat and cool the building. Sometimes, the EUI can be adjusted with respect to heating degree-days (HDDs) and cooling degree-days (CDDs) to partially account for this variation. HDDs, or the average of the daily maximum and minimum temperature subtracted from a base temperature (often 65°F), are often reported as part of the daily weather. Weather normalization is essential in tracking the energy use of a building over time or to compare it to buildings in other locations. Year-to-year variations in weather can impact the energy use of a building up to 15%.

The primary tool used to normalize variables and compare like buildings is the U.S. Department of Energy's CBECS data (EIA 2003). The CBECS provides EUI data across property types and climate zones and reports EUIs in kBtu/ft^2·yr. The ENERGY STAR Portfolio Manager (2008b), in assigning a score of 1 to 100 by building type, uses CBECS data along with other databases. The CBECS information in Table 2-2 provides comparisons of different building types and climate zones. This table is provided for illustrative purposes; it contains too many building sizes and characteristics to be used for a specific building. A climate zone map can be found in Figure 2-2.

Once a building's EUI is calculated and variables are isolated, the result can be compared to any one or all of the following:

- energy history at the facility (e.g., the last several years of energy use)
- comparable properties owned by the organization (e.g., multiple retail outlets)
- similar local facilities (e.g., schools in a state)
- national averages

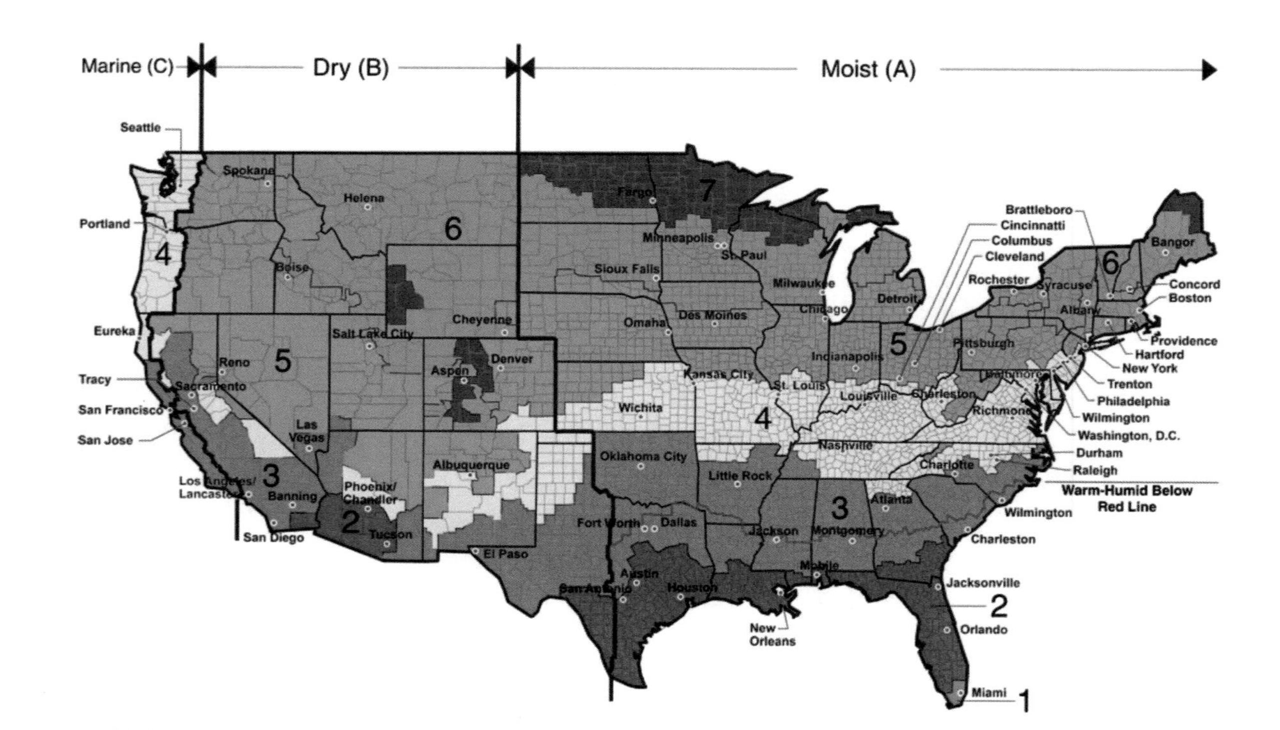

Figure 2-2. Map of the United States displaying U.S Department of Energy climate zones (Briggs et al. 2003).

Table 2-2. 2003 CBECS Weighted Mean Energy Use Intensities by Subsector and Climate Zone: I-P Units (kBtu/ft^2·yr)

Subsectors	Climate Zones														
	All	**1A**	**2A**	**2B**	**3A**	**3B**	**3C**	**4A**	**4B**	**4C**	**5A**	**5B**	**6A**	**6B**	**7**
All	90	74	72	114	89	70	62	95	108	99	104	87	89	97	71
Office/ Professional	93	42	82	72	88	70	58	97	143	95	107	66	110	114	68
Nonrefrigerated Warehouse	42	22	16	—	22	21	20	39	29	37	79	60	37	58	33
Education	83	52	73	160	62	74	105	102	38	58	87	79	90	90	84
Retail (Except Malls)	74	61	93	129	60	50	31	65	100	—	88	80	93	97	102
Public Assembly	94	75	60	—	112	48	45	110	44	249	103	97	88	102	97
Service	77	60	53	—	49	61	27	82	83	—	80	101	88	99	65
Religious Worship	44	—	31	—	28	31	—	47	56	—	52	39	83	34	—
Lodging	94	81	91	—	98	57	—	92	264	545	89	65	108	93	68
Food Services	258	393	208	—	423	393	82	234	—	260	258	228	203	236	192
Inpatient Health Care	249	200	246	360	205	257	204	248	163	—	294	245	240	235	256
Public Order and Safety	116	—	91	—	160	79	—	129	—	—	108	94	126	148	—
Food Sales	200	—	166	—	212	183	120	242	—	—	203	147	242	—	199
Outpatient Health Care	95	19	77	—	55	106	—	70	190	—	111	120	112	91	166
Vacant	21	—	4	47	4	6	0	40	3	60	21	93	22	—	55
Other	79	—	48	—	100	175	—	71	26	—	94	92	69	85	57
Skilled Nursing	125	—	71	—	84	85	—	148	—	—	148	153	118	134	—
Laboratory	305	—	—	—	242	170	—	600	—	—	370	—	268	115	—
Refrigerated Warehouse	99	—	—	—	—	—	—	120	—	—	68	51	62	—	—

Notes:

Values are mean EUIs weighted by floor area

Climate zones are defined in *ANSI/ASHRAE Standard 169-2006, Weather Data for Building Design* (ASHRAE 2006b). Climate zones with no data had no samples in the 2003 CBECS. For missing climate zones, use EUIs for the most similar climate zone (e.g., zone 7 for 8 or zone 2A for 2B).

- building code or best practices standards (i.e., as a percentage of *ANSI/ ASHRAE/IESNA Standard 90.1, Energy Standard for Buildings Except Low-Rise Residential Buildings* [ASHRAE 2007d]).

TYPES OF RESULT COMPARISONS

This section discusses self-referencing (historical comparisons), internal portfolio referencing (comparisons within a portfolio), and external referencing (metric comparison to averages and multi-parameter comparisons through ENERGY STAR) methods. The appropriate measurement method depends on the type of facility, type of data available, and availability of meaningful performance metrics.

Self-Referencing Historical Comparisons

The self-referencing comparison is a review of current performance against historical performance. It requires only data regarding the facility of interest. In order to isolate the energy performance of the facility, factors such as weather and change of use need to be considered in the comparison. A historical comparison can also allow for the assessment of subsystems (e.g., the cooling system, lighting, fans, etc.) within a building if sufficiently detailed energy use data is available. One example of this type of assessment uses a correlation of past electricity and natural gas use to outdoor temperature to establish a baseline. Ongoing reviews of performance (e.g., daily energy use compared to the trend) can then be used to signal operators about potential issues when performance deviates from past performance.

The historical performance review is useful in tracking performance and showing the impact of changes to the facility. It does not, however, answer the question of how well a building's performance compares to that of other facilities. In undertaking these performance reviews, operators should ensure that the building systems (e.g., temperature setpoints, outdoor air ventilation control, etc.) are operating as intended.

Internal Portfolio Referencing Comparisons

An internal reference is possible when an organization has multiple facilities that share a similar function. Retail stores, supermarkets, and schools are all examples. If these facilities are in geographically similar regions, there may not be a need to normalize for weather, as all operate under the same conditions. A simple comparison of how performance metrics rank can provide insight into which facilities are consistently performing better. The goal is often to try to shrink the difference between the best and worst performers by improving the facilities with the lowest performance metrics. As with historical performance evaluations, it is difficult to compare the internal reference with facilities outside of the portfolio. The best-performing facility in a portfolio might not compare well with a similar facility belonging to another organization or compare well with building codes or energy standards—such as those set forth in ASHRAE/IESNA Standard 90.1 (ASHRAE 2007d).

Weather normalization and base/process load adjustment calculations can be simple or complex, depending upon the nature of the facility. Simpler normalizations should be utilized whenever possible, as it is important to establish a tracking system that can be maintained over time.

External Referencing Comparisons

Using the CBECS, a facility's energy performance can be compared to an array of similar buildings. For example, an office building can be compared to a statistical population of office buildings to determine how efficient the office building is in comparison to other office buildings. The external reference provides a greater perspective for performance comparisons.

Tables 2-3 and 2-4 provide examples of EUI data sets for specific building types. The first EUI data set example (Table 2-3) combines electricity and fossil fuel use by converting both to Btus based on a simple unit conversion of kWh to Btus (ASHRAE 2007a). This is done without considering source energy use to create electricity or losses in transmission. (Note: site and source energy are described later in this chapter.) The table provides the range in EUIs observed through the CBECS survey. The 20,000 ft^2 office building presented previously had an EUI of 96,560 $Btu/ft^2 \cdot yr$. It would be just above the 75th percentile in Table 2-3, meaning that three quarters of the building population of similar buildings is more efficient than the example.

The second example (Table 2-4) shows only the electricity use on a gross floor area basis with values across the distribution of observations (ASHRAE 2007a). The example building, with 5 kWh/yr per gross ft^2, would have a below-average electric EUI between the 10th and 25th percentile. Its high energy use is a result of high fossil fuel energy use.

METHODS FOR TRACKING AN EUI

Building operations and occupant use of buildings vary over time. Therefore, energy use must be tracked over time, and the indices by which energy use is rated must be redetermined on a periodic basis. There are a number of computer-based tools that will carry out much of the process described above. The most prominent and widely used is EPA's ENERGY STAR Portfolio Manager (2008b). ENERGY STAR Portfolio Manager is free and can be accessed at www.energystar.gov/index.cfm?c=evaluate_ performance.bus_portfoliomanager. (The next section provides more information about this tool.)

On the commercial side, there is a genre of software referred to as *energy accounting packages*. These were some of the earliest programs available to assist

Table 2-3. 2003 Commercial Sector Floor Area and EUI Percentiles

Building Use	Calculated, Weighted		Actual Number of Buildings, *N*	Calculated, Weighted EUI Values Site Energy, kBtu/yr per gross ft^2					
				Percentiles					
	Number of Buildings, (100s)	Floor Area, 10^9 ft^2		10th	25th	50th	75th	90th	Mean
Administrative/ Professional Office	442	6.63	555	28.1	41	62	93	138	75
Bank/ Other Financial	104	1.10	75	55.7	67	87	117	184	106
Clinic/Other Outpatient Health	66	0.75	100	28.7	41	66	97	175	84
College/ University	34	1.42	88	14.1	67	108	178	215	122

Source: ASHRAE 2007a

Table 2-4. Electricity Index Percentiles from 2003 Commercial Survey

Building Use	Weighted EUI Values, kWh/yr per gross ft^2					
	Percentiles					
	10th	25th	50th	75th	90th	Mean
Administrative/ Professional Office	3.54	6.7	11.0	15.0	24.1	12.7
Bank/ Other Financial	6.23	14.5	22.2	29.5	27.3	16.6
Clinic/ Other Outpatient Health	4.94	9.4	15.2	20.7	27.3	16.6
College/ University	4.13	10.5	15.0	24.0	42.3	17.7

Source: ASHRAE 2007a

professionals in making facilities more energy cost efficient. Energy accounting packages are available that are quite user friendly and can be used by both facility owners and managers and by energy and engineering professionals. Some of the major software packages on the market can be reviewed at the California Energy Commission Web site in their report listing of efficiency handbooks at www.energy.ca.gov/reports/efficiency_handbooks/. All of these applications will take in historical and ongoing energy use and cost data, as well as facility operational information (e.g., hours of operation, occupancy, loads, etc.), and through a series of automated computations compute the EUI of a building. The more sophisticated of these applications will produce monthly energy use report cards comparing a building's energy use and costs to those in a previous period as well as compare all buildings within an organization's portfolio to one another.

ENERGY STAR PORTFOLIO MANAGER

In an effort to evaluate energy consumption patterns of commercial buildings, the EPA created the ENERGY STAR Portfolio Manager program (2008b). Powered by energy analysis tools, this program can aid building and facility managers in rating the energy performance of buildings, comparing building consumption patterns to similar buildings, and prioritizing energy efficiency investments. Top-performing buildings (achieving a score of 75 or more and validated by a professional engineer) can earn the ENERGY STAR label. This program is also used to meet the USGBC's energy efficiency criteria for the Leadership in Energy and Environmental Design (LEED) for Existing Buildings label. The ENERGY STAR label lasts for one year. Good energy management and continued tracking of building energy use ensures that buildings continue to operate in the top quartile of energy efficiency. Additionally, to earn the label, the building's 12-month average occupancy must be 75% or greater. Organizations that show a 10%, 20%, 30%, or more reduction in normalized energy use or achieve a 75 rating average across building portfolios can earn recognition as ENERGY STAR Leaders. A high score is indicative of good energy management practices and can be used to market leasable space.

Methodology

The ENERGY STAR Portfolio Manager (2008b) facilitates the comparison of a building to the national building stock, while accounting for building type, use, and weather. The tool uses correlations of energy use to building characteristics to adjust for typical parameters such as hours of use and level of occupancy. Each building type has specific building operation parameters that are used to determine predicted energy use, such as

- percent of building that is air conditioned,
- amount of outdoor air to be conditioned,
- number of computers for a school, and
- number of beds for a hospital.

Building Types that Can Achieve the ENERGY STAR Label

At this time, the ENERGY STAR label and the assignment of a score from 1 to 100 is only available for the following building types:

- bank/financial institutions
- courthouses
- hospitals
- hotels and motels
- K-12 schools
- medical offices
- offices
- residence halls/dormitories
- retail stores
- supermarkets
- warehouses

The Portfolio Manager requires the user to enter energy use data covering at least 11 consecutive months and building characteristic data, which varies with the building type. Floor area and weekly hours of use are common characteristic parameters used in most building types.

The tool produces a score between 1 and 100 that represents where the building falls in the distribution of buildings with similar characteristics. A score of 50 designates that 50% of similar buildings use more energy and 50% use less energy. A score of 75 indicates the building is performing in the top quartile of similar buildings.

Other Building Types

For building types that are not included in the benchmarking tool, the EPA has developed tables showing typical energy use for a variety of other buildings based on CBECS data:

- colleges/universities
- convenience stores
- restaurants/cafeterias
- quick-serve restaurants
- lodging facilities
- shopping malls
- nursing homes/assisted living facilities

Space and utility data for all property types can be loaded into the ENERGY STAR Portfolio Manager (EPA 2008b) and the program will track weather-normalized source energy use over time.

The *2008 Professional Engineer's Guide to the ENERGY STAR Label for Commercial Buildings* is available on the ENERGY STAR Tools & Resources web page and can be downloaded at www.energystar.gov/ia/business/evaluate_performance/pm_pe_guide.pdf (2008a).

The Portfolio Manager tool (EPA 2008b) allows users to aggregate property data to create a picture of an entire portfolio's performance. After establishing a baseline rating, the program allows users to record capital expenditures, track portfolio-wide improvements, and monitor progress toward energy-saving goals. Additionally, the Portfolio Manager enables users to monitor the carbon emissions associated with energy consumption as well as emissions avoided due to energy performance improvements. All of the portfolio information in the tool may be organized into customized views that match a company's management structure, geographical structure, or any other logical arrangement.

BUILDING RENOVATIONS

While ENERGY STAR Portfolio Manger (EPA 2008b) is well-suited for occupied buildings, buildings undergoing significant building renovations should refer to the Advanced Energy Design Guides (AEDGs). To determine whether building systems are properly sized and how well the building will perform once occupied, the building is modeled to building codes and standards such as ASHRAE/IESNA Standard 90.1 (ASHRAE 2007d). High-performance buildings are 15%–40% more efficient than the minimum requirements set forth in ASHRAE/IESNA Standard 90.1. The AEDGs are meant to be used with smaller buildings and selected building types. The documents are available for free download or hard copy purchase from www.ashrae.org/aedg. For large buildings, or those not covered by an AEDG, the building's anticipated energy performance would be compared to ASHRAE/IESNA Standard 90.1. ENERGY STAR has a program called Target Finder that can be used to see how a building's modeled energy consumption ranks within ENERGY STAR.

SETTING ENERGY TARGETS

Energy or carbon reduction targets are set relative to some measure of performance in the population or some historic level of performance. Once the energy performance of a building is established, such as an ENERGY STAR score of 65 or an EUI of 165,700 Btu/ft^2·yr, energy targets can be set based on the goals of the company and asset strategy. Based on the data sources discussed in the chapter, examples of various energy targets follow:

- improve 10%, 20%, 30%, or more beyond historical performance
- improve outliers (buildings with larger-than-expected EUIs) in a portfolio of similar buildings
- reduce by 20% carbon emissions resulting from real estate by 2010

- earn the ENERGY STAR Partner of the Year Award for ENERGY STAR Leaders
- qualify the building each year for the ENERGY STAR label (an ENERGY STAR score of 75 or higher)
- pursue LEED for Existing Buildings: Operations & Maintenance certification (a minimum ENERGY STAR score of 69 or being in the 19th percentile of comparable building or higher)
- consider demonstrating performance above an existing energy standard when the building is undergoing a major rehabilitation such as replacement of the entire HVAC system, etc.—typically mandatory under local codes

Target setting depends on the data available, the methods available for measuring performance, and the commitment of management to providing both the means of creating and tracking performance metrics and making improvements in operational practices and possibly capital improvements. Baselining energy performance and continually tracking progress toward defined goals provides the surest means to measure improvements in energy efficiency.

ENERGY VALUE OF SOURCE VS. SITE ENERGY

The production, transportation, and long-distance delivery of energy through existing pipes, wires, and delivery vehicles have various levels of loss and inefficiency, which are for the most part already minimized and unavoidable based on currently available and/or feasible technology and operations. Large power generation facilities, particularly older boiler-turbine-based facilities, will convert approximately one-third of the energy they consume into electrical power that is actually delivered to the electrical grid. Newer facilities are achieving a better than 50% conversion, but this still means half of the energy in whatever fuel consumed never reaches the grid as usable energy.

This loss continues as the electrical power flows through the transmission and distribution systems, losing another 10% or more of the energy that never reaches the end user. Each kWh delivered to a facility may have required 3 to 4 kWh of energy production at the source. Natural gas and district steam delivery systems also realize losses from point of production to end user. Fuel oil delivered to a facility for its boilers has a loss factor of about 0.5% in the fuel and other costs used to transport, store, and blend it before it is delivered.

The end result is that every kW or Btu of energy saved (or produced) on-site can have a much higher equivalent value in reduced energy output at the source. The carbon emissions associated with a building are not only impacted by its direct energy consumption but also the energy consumed or lost in the production, transportation, and delivery of energy to the building. As an example, site-source ratios were used in the recently published *Emissions Inventory for New York City* (NYC 2009) to determine the actual emissions footprint of energy usage in buildings within the city. One such ratio showed that more than 1800 Btu of fuel energy input is required to deliver 1000 Btu of energy to a facility through the local district steam system.

SOURCE AND SITE ENERGY USE

One of the main difficulties associated with comparing energy use among buildings is comparing buildings that use different forms and sources of energy. A source EUI is one means to combine energy types such as electricity and natural gas. As opposed to a site EUI, a source EUI attempts to normalize varied fuel mixes utilized at each specific building so that it can be compared more readily to a national average performance.

Before site energy consumption can be translated into source energy consumption, the origin of the utilized energy, either heat or electricity, needs to be understood. In particular, energy acquisition may be from primary or secondary sources. Depending on the operational characteristics of the building, heat and/or electricity may be produced on site (via boiler or microturbine). Otherwise, the building owner or manager may purchase power from a utility or heat from a district steam service. *Primary energy* refers to the raw fuel used to produce this heat or electricity while *secondary energy* refers to the product of generation, namely heat and electricity.

To convert primary energy use into source consumption, any losses associated with storage, transportation, and delivery of the fuel to the building need to be accounted for. Source EUIs are created by multiplying site EUIs by a source-site ratio (the source-site ratio is the source energy/usable energy available at the site). The final result represents the primary and secondary energy in terms of the total equivalent source energy. (See Table 2-5.)

One source for the source-site ratio is a normalized national average computed by a methodology developed by the EPA. It would be expected that different buildings and utilities would have different energy conversion efficiencies, and the EPA addresses this situation by using these national source-site ratios that are specific to each raw fuel type.

The ENERGY STAR Portfolio Manager (EPA 2008b) takes into consideration site and source energy in calculating a building's energy efficiency. In this way, building operators are not penalized for the relative inefficiency of a local utility. Conversely, buildings with very efficient on-site equipment will be rewarded (EPA 2008c). Refer to ASHRAE Standard 105-2007 (ASHRAE 2007e) for more information on source vs. site energy.

The ultimate merit of source EUIs is the ability to compare a specific building's energy use to that of other buildings on a basis that accounts for the total impact of the building's energy use.

Table 2-5. Source-Site Ratios for all Portfolio Manager Fuels

Fuel Type	Source-Site Ratio
Electricity	3.340
Natural Gas	1.047
Fuel Oil (1,2,3,4,5,6, Diesel, Kerosene)	1.01
Propane and Liquid Propane	1.01
Steam	1.45
Hot Water	1.35
Chilled Water	1.05
Wood	1.0
Coal/Coke	1.0
Other	1.0

Source: EPA 2008b

Improving Energy Performance 3

INTRODUCTION

Once the energy efficiency of a building has been determined and energy performance goals have been established, the most cost-effective measures for improving energy performance should be determined. Energy in a building can be characterized by the end uses or functions the energy serves (e.g., lighting, plug loads, building envelope, distribution of heating and cooling, water heating, refrigeration, and heating and cooling equipment). Immediate energy savings can be recognized through operational adjustments and occupant behavior changes. Energy efficiency measures (EEMs) are most often identified through energy audits and energy-focused retrocommissioning. Of paramount importance in implementing EEMs is taking an integrated approach when evaluating building systems and staging building improvements. An energy efficiency plan is the natural outgrowth of these activities.

WRITING AN ENERGY EFFICIENCY PLAN

Once the building energy use has been determined and a list of energy conservation measures has been identified, it is time to write the energy efficiency plan. The plan should be action-oriented. It should consist of an existing building's energy utilization index (EUI) and energy cost index, savings goals, and a list of energy-saving measures. It should also contain timelines, budgets, and follow-up activities to ensure that savings are obtained and continue to be realized over time. The plan

- allows building owners and operators to track progress,
- permits financial planning for capital improvements or replacements, and
- permits owners to have contingency plans to take advantage of unforeseen circumstances such as unexpected equipment failure or changes in tenants.

The plan should be developed in conjunction with operating staff to obtain their buy-in. In a successful program, staff will understand what to do and how to do it, and they will understand that the new procedures make their jobs easier. Examples that need be presented in the plan include the following:

- fewer complaints from improper space comfort conditioning
- reduced occurrences of unexpected equipment failure
- pride that comes from being responsible for a well-oiled machine (building)

TAKING AN INTEGRATED APPROACH TO IMPROVING ENERGY EFFICIENCY

To achieve cost-effectiveness and energy efficiency, it's necessary to take an integrated approach to improving energy performance, meaning that the building as a whole system needs to be examined to identify ways in which individual building systems can enhance one another. For example, building heating, ventilating, and air-conditioning (HVAC) systems at original installation are sized based on maximum expected capacity, which is based on anticipated occupancy and weather extremes, usually with a healthy safety factor. (See Figures 3-1 and 3-2.) The amount of cooling the building actually needs in practice may be less and is impacted by many factors, including heat generated by the type of lighting, office equipment, and appliances in place. New energy-efficient lighting and equipment produce less heat, requiring less cooling. Another example is that the building's envelope and its ability to protect the thermal comfort of the indoor environment play a significant role in determining cooling needs. Thus, replacing a black roof with a white or vegetated roof will significantly reduce cooling requirements, particularly in places like Arizona, Florida, and Texas. Therefore, EEMs affecting the building's cooling needs such as lighting, plug load, and envelope should be performed first. This will allow for the right sizing of HVAC equipment.

Improving energy performance is not limited to demand reduction or using less. Strategies for improving efficiency include a) using free energy such as daylight and outside air, b) recovering waste energy by using the heat generated by equipment for heating, and c) increasing the efficiency of existing systems such as lighting, envelope, and HVAC equipment. The benefits of an integrated approach apply to the entire life cycle of a building (i.e., construction, alterations and improvements, renovations, retrofits, and ongoing operations and maintenance).

No- and Low-Cost Operational Adjustments

Opportunities to save energy without sacrificing comfort often exist by making operational adjustments and implementing energy management prac-

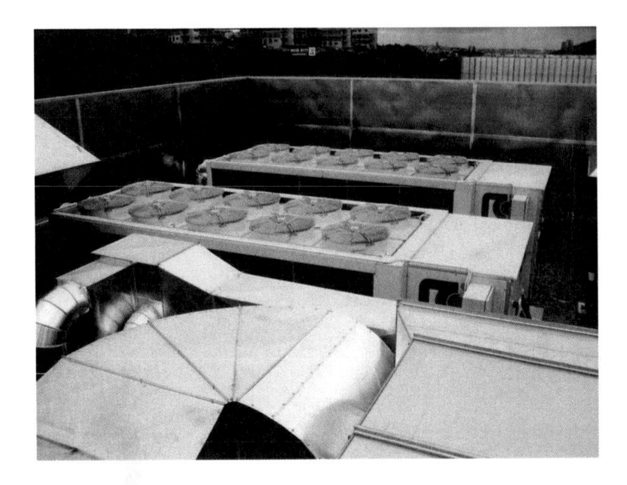

Figure 3-1. Rooftop chillers are less efficient than water-cooled units, but their impact on energy efficiency can be minimized when used in moderate summer conditions and depending on the design strategy.
(Photograph courtesy of *High Performing Buildings* [Parsley and Serra 2009].)

Figure 3-2. Small mini-split systems can improve energy efficiency in less-occupied support areas.
(Photograph courtesy of *High Performing Buildings* [Schreiber 2009].)

tices in operations and maintenance (O&M). Many actions that reduce energy use may also improve the comfort, satisfaction, and productivity of occupants if properly implemented. Facility operating staff can contribute to building energy efficiency by implementing low-cost actions and monitoring performance, especially when such actions become part of organizational policies. Operational adjustments include minimizing HVAC system operating hours by examining operating hours and indoor air quality (i.e., starting up and shutting down systems based on occupancy). An example of an operational adjustment that brings meaningful savings is changing the building practice of always running the HVAC system on weekends, as posted in most leases, to running it on weekends by request only. Since tenants directly or indirectly pay the utility bills, this EEM is being widely embraced by both tenants and landlords. Implementing O&M programs targeting energy efficiency can save 5% to 20% on energy bills without a significant capital investment. O&M energy management best practices can be found in Chapter 5.

Change Occupant Behavior and Institute Purchasing Criteria

An easy way to reduce plug load is to establish minimum energy standards for purchases—within an organization and within a multitenant office building. Increasingly, landlords are requiring through lease language only ENERGY STAR-rated appliances, compact fluorescent lights (CFLs), and office equipment.

Energy savings of 3.5% to 15.2% can be recognized through changes in occupant behavior according to the Building Owners and Managers Association (BOMA) International Building Energy Efficiency Program (BEEP) (2006a). EEMs may include the following:

- *Appliances and Food Service:* Allow only ENERGY STAR-labeled appliances (e.g., refrigerators, dishwashers, washers, etc.) and commercial food service equipment during new installations.
- *Office Equipment and Electronics*: Rising energy consumption in the office sector can be attributed to the use of more computers and office equipment. Reduce plug load by purchasing only Electronic Product Environmental Assessment Tool (EPEAT) (www.epeat.net) and ENERGY STAR-rated (www.energystar.gov) office equipment (e.g., computers, printers, monitors, fax machines, copiers, etc.) and ENERGY STAR-rated electronics (e.g., TVs, DVD players, and water coolers) and installing power management software. Electronic equipment can be controlled manually through the use of power strips that can be switched off by the occupant or by occupancy sensors.
- *Energy Awareness Program:* For EEMs that are not controlled electronically, institute an energy awareness program for occupants that includes reminders for turning off equipment and coffee pots when not in use. Also, close south-facing blinds and

use task lighting instead of area lighting. ENERGY STAR has developed a program called Bring Your Green to Work that gives tenants and occupants suggestions for how to reduce energy use in the office. Information is available at www.energystar.gov/index.cfm?fuseaction=bygtw.showSplash.

To assist in the development of energy standards for purchasing, the government agency Web sites below may be useful resources:

- The U.S. Department of Energy (DOE) Federal Energy Management Program's Energy-Efficient Products Web page: www1.eere.energy.gov/femp/procurement/index.html
- Office of the Federal Environmental Executive's Green Purchasing Web page: http://ofee.gov/gp/gp.asp
- The ENERGY STAR Quantity Quotes site for connecting purchasing groups with suppliers for bulk purchasing of ENERGY STAR-qualified products: www.quantityquotes.net/default.aspx

ANALYZING HOW AND WHERE ENERGY IS USED

Beyond operations and maintenance energy management best practices, managing plug loads with energy-efficient electronics, implementing an occupant awareness program, and identifying EEMs and their potential values requires an analysis of how and where the building uses energy. This challenging and complex work (referred to as *energy audits* and/or *energy-focused retrocommissioning,* as appropriate), particularly in larger or high-rise buildings, requires experienced engineers.

Energy Audits

While the purpose of an energy audit is consistent (i.e., to identify ways to save energy), the scope of energy audits can vary widely. It is generally agreed that three levels of energy audits exist—walk-through, survey and analysis, and investment grade. However, the resulting end product of each of these can vary widely without a defined scope. Using a defined scope for energy audits provides consistency when it comes to quality, reporting and credibility (which facilitates analysis), bulk-bidding, and financing. The Leadership in Energy and Environmental Design (LEED) for Existing Buildings: Operations & Maintenance Rating System uses American Society for Heating, Refrigerating and Air-Conditioning Engineers, Inc. (ASHRAE) energy audit scopes as the prescriptive for meeting program requirements. The energy audit scopes and accompanying tools can be found on page five of *Procedures for Commercial Building Energy Audits*. The following summaries were taken from that page (ASHRAE 2004b).

- **Level I—Walk-Through Analysis:** A Level I energy analysis will identify and provide a savings and cost analysis of low-cost/no-cost measures. It will also provide a list of potential capital improvements that merit further consideration and an initial judgment of potential costs and savings.

- **Level II—Energy Survey and Engineering Analysis:** A Level II energy analysis includes a more detailed building survey and energy analysis. A breakdown of energy use within the building is provided. This energy analysis will identify and provide the savings and cost analysis of all practical measures that meet the owner's constraints and economic criteria, along with a discussion of any changes to O&M procedures. It may also provide a listing of potential capital-intensive improvements that require more thorough data collection and engineering analysis and a judgment of potential costs and savings. This level of analysis will be adequate for most buildings and measures.
- **Level III—Detailed Analysis of Capital-Intensive Modifications:** A Level III energy analysis focuses on potential capital-intensive projects identified during the Level II analysis and involves more detailed field data gathering as well as a more rigorous engineering analysis. It provides detailed project cost and savings calculations with a high level of confidence sufficient for major capital investment decisions.

An energy auditor may be required for Level II or Level III audits, depending upon the skill and workload of in-house staff. An experienced energy auditor brings the perspective of having performed energy audits in a variety of buildings.

The auditor will often employ some survey and monitoring test equipment/tools (e.g., thermometers, light meters, combustion efficiency kits, etc.). The more information gathered on a building and its energy use process, the better the resulting analysis. The auditor is a forensic detective who will be combining the clues from the data compiled in the pre-site visits with what he or she observes, sees, or hears during the on-site interviews with operators and occupants, and with the data that has been collected, from nameplates or with monitoring equipment, during the visits. Combining all of this information and using a very critical experience filter enables the auditor to determine the potential opportunities to reduce energy costs for the facility. The site visit should also serve to educate the management and operators about their facility's equipment and what they need to do to get the most out of it. Unfortunately, this educational part of the audit often gets overlooked.

Once all the data has been gathered, the auditor analyzes energy use and demand, segregating consumption into baseload, weather-dependent load variations, and any potential major service and/or process loads. The auditor will then adjust for parameters such as weather, production, occupancy, and sales (in the case of retail establishments). Consideration will be given to building system and measure interactions when evaluating energy efficient measure (EEM) opportunities so as not to double count savings. There should also be a focus on operations and management (O&M) or management measures. These are often the largest opportunity for savings, and they help ensure savings over time. In considering the value of a set of EEMs, facility management should insist that the auditor make proper use of applicable utility rates, usually those on which the purchased energy is based. The savings or cost reductions can only be from the marginal parts of the applicable rate block components. In other words, savings are not based upon the average cost of a unit of energy but on the last unit of energy purchased. This is important since the price of energy often decreases as more energy is purchased. Additionally, factors such as season and time of use (or day) or ratchet implications should be taken into account. This ensures that the actual financial returns are properly represented. The returns on making the improvements are evaluated by looking at the cost of the improvements—from material and installation labor to disposal of existing equipment, engineering design (if necessary), permits and fees where applicable, and other life-cycle costing (LCC) items. (LCC is discussed in Chapter 4.)

Once all of the analyses have been completed and the EEMs with economic effectiveness greater than the hurdle rate set by the user (or return on investment level) have been identified, an energy audit report is prepared to communicate the findings. Well-written reports will become action plans for both the short and long term. Reports should be prepared and presented in a manner that will allow for use by building management and operating personnel as the blueprint to get and keep the building more energy efficient. These reports can often be a part of the application process for funding sources (such as when loans or grants are applied for from banks or government agencies) to ensure that the analyses are cogent and can be relied upon to provide the financial returns projected. A good report also acts as a reference document for contractors' actions (in implementing the EEMs) and for site management in dealing with future problems.

The report should clearly delineate any assumptions and parameters used in the analysis to justify the recommendations. This becomes particularly important when going for funding to pay for measures, as the lending institution will often have their engineers review the audit. This allows reviews to occur without slowing down the capital procurement process.

The report should also make recommendations for broader preventative maintenance routines and/or management programs such as a steam trap-testing program or a heating, ventilating, and air-conditioning (HVAC) system maintenance log. This should be done to ensure persistence of savings, equipment longevity, and occupant comfort. The organization and presentation of an energy audit report varies, but a good report does an effective job in communicating the findings while not overwhelming the reader with volume for the sake of volume. The main areas will be an executive summary, a description of conditions and recommendations, and appendices (for detail on those items introduced earlier in the report). The report should also present enough information to ensure that future design meets the intent of the auditor's recommendations and can achieve the savings that were projected. A simple example of that would be: rather than suggesting to “replace all windows,” say “replace windows with double pane, low-E, argon filled.” Lastly, the auditor should make himself or herself available to meet with the facility management to review the recommendations and do any necessary follow up.

A value-added service that can be provided by the auditor is to provide measurement and verification (M&V) options. These can be utilized to measure the performance of systems and verify the savings or reduction in energy use or lowered demand. This takes advantage of the time the auditor has spent thinking about how the building operates and the ways in which the EEMs will affect that operation. M&V is valuable in larger organizations as it provides the ability to demonstrate to upper management that the money spent on EEMs at a site provided the returns expected. When working with an energy services company (ESCO) under a performance contract, a portion of the savings is often paid to the ESCO, and M&V assures the facility management that the savings have occurred. The International Performance Measurement and Verification Protocol provides guidance in establishing M&V plans. *ASHRAE Guideline 14-2002, Measurement of Energy and Demand Savings* (ASHRAE 2002) provides more technical information on implementing an M&V plan and calculating savings.

With any level of energy audit, it should be understood that any engineering work required to design and/or implement a measure will be a separate service and should not be expected to be part of the same scope of work as the audit. This helps ensure that the auditor, when determining the recommendations, is focused on what is best for the building, not what will add additional hours to this project. Related to this, the auditor may not necessarily be the most qualified to do such design work, as his or her specialty should be the forensic engineering required to ascertain the available opportunities that exist at a site.

Some firms market a product called an *investment grade energy audit* (IGA). This is a term that is used to sell a very wide range of services, so buyers should beware when considering them. Those services offered under the label of IGA represent analyses of a highly varied level of rigor with regard to technical and economic evaluation. The level of service provided in an IGA cannot be assumed to be the same or necessarily the best level of service in all cases.

Commissioning

Studies consistently reveal that buildings more often than not do not operate as they were designed, which directly impacts energy efficiency and occupant satisfaction. Commissioning is the process of determining whether a building is operating as it was designed and taking corrective action if it is not. Existing building commissioning is called *retrocommissioning* and involves developing a building operating plan that identifies current building operating needs and requirements and then testing building systems to determine whether they are operating as planned and optimally. *Recommissioning* is applying commissioning to a project that has previously been commissioned or retrocommissioned. *Ongoing commissioning* incorporates ongoing measurement and verification to ensure the building continues to operate as commissioned. Ongoing commissioning is used to resolve operating problems, improve comfort, and optimize energy use. Because this Guide pertains to existing buildings, the focus of this section will remain on retrocommissioning.

Retrocommissioning. Much like the scope of an energy audit, the scope or definition of *retrocommissioning* can vary widely. To commission a building is to compare it to its intended design potential. Existing building systems often don't operate as they were designed to. Existing buildings, particularly multitenant office buildings, experience turnover—from tenant build-outs to management and ownership changes. Therefore, retrocommissioning an existing building requires establishing how the building should be currently operating to maximize energy efficiency, occupant comfort, air quality, and equipment life span. Commissioning with a focus on energy efficiency can produce a different result than the original design parameters. However, the intent is to improve energy efficiency without compromising occupant comfort and indoor air quality.

The retrocommissioning process begins with an investigation phase to understand how building systems are currently operating and are maintained. Then, EEMs are identified, and a plan is developed for the most cost-effective implementation of the EEMs. The purpose of the investigation phase is to determine how the building is currently operating and how efficiency could be improved. To organize this process, the LEED for Existing Buildings: Operations & Maintenance Rating System requires, as a prerequisite, that each building establish the following: a) sequence of operations, b) operating plan, c) system narrative, and d) preventive maintenance program. *ASHRAE Guideline 0-2005, The Commissioning Process* (ASHRAE 2005b) provides a detailed description of each of these requirements. These components are also part of the systems manual discussed in Informative Annex O of ASHRAE Guideline 0. In summary, these building documents are integral to an energy-efficiency-focused operations and maintenance program, which is discussed further in Chapter 4.

The Cost/Payback of Retrocommissioning. Retrocommissioning is one of the most cost-effective means of improving energy efficiency in commercial buildings. A study by Lawrence Berkeley National Laboratory (LBNL) (Mills et al. 2004) included 224 new and existing buildings that totaled more than 30 million ft^2 of commissioned floorspace (73% existing buildings and 27% new construction) and included office, retail, hotel, education, laboratory, and hospital commissioning projects throughout the United States. The findings for existing buildings were as follows:

- The median cost to fully commission an existing building is \$0.27/ft^2 of floor space.

- The average resulting energy savings per building was 15%.
- The average time to pay back the cost of commissioning per building was 0.7 years.

Selecting Buildings for Audits And Retrocommissioning

While all buildings can be served by an energy audit, the payback and benefits are greater for some than for others. Buildings performing below the top quartile (i.e., with less than a 75 ENERGY STAR score) are good candidates for energy audits. The lower the ENERGY STAR rating or the higher the energy utilization index (EUI), the greater the need for an energy audit and the greater the potential payback. Buildings that are best suited for retrocommissioning have one or more of the following attributes:

- They need major upgrades. In this case, the investigative phase of retrocommissioning may precede design and construction, and commissioning will occur after construction.
- They are newer buildings with complex mechanical and energy management systems that have not been commissioned and/or have never worked properly.
- They have undergone many space/use reconfigurations.
- They are experiencing HVAC issues that are leading to recurring occupant complaints about thermal comfort.

Regardless of the method of identifying EEMs, it is important ensure that a systematic, integrated approach is taken in developing the plan that is implemented. Additionally, energy audits and retrocommissioning are more attractive when utility rates are high and if energy audit and retrocommissioning rebates are available through the utility company.

The EEMs identified by an energy audit or retrocommissioning will most likely include O&M adjustments, occupant behavioral changes, and system changes. The latter often require some form of capital investment. While Chapter 4 will discuss the economic analysis of these investments, following sections are an introduction to the most common energy end uses or systems within a building.

CONSIDERING ENERGY USAGE BY END USE

To determine the greatest areas of opportunity for energy savings, the energy the building uses must be broken down into its end uses. According to the Energy Information Administration (EIA) (2006), there are 4.9 million commercial buildings containing 71.6 billion ft^2 of floor area. These buildings consumed 18 quads or 18% of the national primary energy use in 2006. The energy use is further refined through end use estimates, as shown in Figures 3-3a and 3-3b. Space heating and cooling dominate, closely followed by lighting.

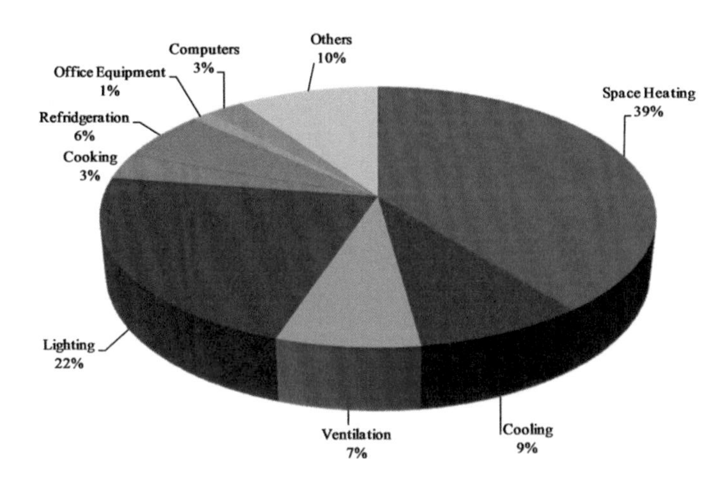

Figure 3-3a. Commercial building energy end-use estimates for heating-dominated climates (EIA 2003).

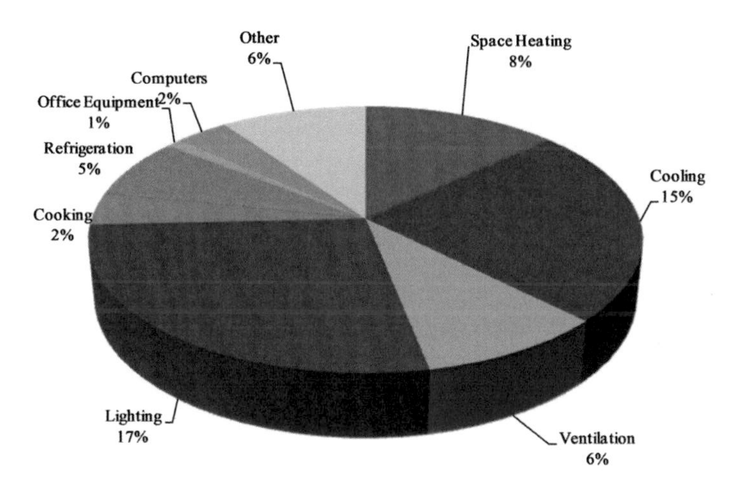

Figure 3-3b. Commerical building energy end-use estimates for cooling-dominated climates (EIA 2003).

Various building types have different distributions of energy use. For example, in retail, less energy is used for office equipment, but space heating and cooling are still predominant. A breakdown of end-use estimates by building types can be found on EIA's Web site at www.eia.doe.gov/emeu/cbecs/cbecs2003/.

Understanding where a building uses energy helps prioritize efforts for improving energy performance and offers insight about where to look for improvements. The following sections provide an overview of the main energy uses in commercial buildings and identify some potential efficiency improvement opportunities. It must be stressed that these improvements are general in nature; the best energy savings opportunities may be unique to a specific building.

Lighting

The lighting systems of a building have both an efficiency level and an operating schedule that define the energy use. The *efficiency* refers to the technology of the lighting system: lamps, ballasts, and fixtures. As technology has advanced, many facilities have undertaken lighting upgrades that include new, more efficient components or entire systems. The energy use of lighting systems has been so high that capital projects often have attractive rates of return.

Just as important as efficiency is how the lights are used. Systems with lighting controls that allow the energy use to track the operation of the building have the potential for the lowest energy use. Occupancy sensors, multilevel switching, and automatic scheduling are all examples of controls designed to use lighting energy only when needed by the occupants. Personal dimmer controls can also save energy and improve occupant satisfaction. Savings of 10% to 20% are typical. Controls are also available to vary the amount of electric lighting used in response to the amount of daylight entering the space. Natural lighting can provide significant energy savings in the perimeter areas of buildings.

Lighting serves utilitarian as well as aesthetic needs. The quantity and quality of light are important in most cases. Understanding and defining the important characteristics is vital before making changes for energy savings. Lighting color, glare, diffusion, contrast, luminance, and fixture style are all elements of varying importance in each lighting application. If major changes are contemplated, a good lighting designer can provide a system that is functional and efficient and provides good aesthetics.

Lighting energy use is mostly correlated with a building's occupancy schedule. Some daily modulation occurs with photocell control of light levels in daylighting systems, and some seasonal variation occurs with photocell control of exterior lighting due to variations in daylight hours. Some amount of lighting energy would be expected to be used continuously for emergency lighting. Quantifying this amount is important to determine if additional lighting is being used during unoccupied periods (e.g., during cleaning).

Lighting technologies have improved in the past ten years and there are many cost-effective improvements that can be made. Light-emitting diode (LED) exit signs use 2 to 4 W of electricity and last 20 years. Labor savings from a decrease in frequency of relamping is as valuable as the energy savings.

In most applications, incandescent lamps should be replaced with compact fluorescent ones. They combine energy savings and longer lamp life. Lamps are available in a variety of color temperatures, not just the cool white lamps that were the only type available just a few years ago. The proper combination of lamp and reflector will produce an appearance similar to an incandescent lamp. CFLs vary widely in quality, so if selecting CFLs in-house, test a sample before buying large quantities.

Super T-8 fluorescents are available in a variety of wattages. For example, a high-efficiency super T-8 ballast with two 28 W super T-8 lamps produces the same amount of light as a magnetic T-12 ballast with two 34 W lamps, while using only 60% of the power and eliminating flickering. Other super T-8 choices use less power and produce less light. Figure 3-4 illustrates the effects of different lighting strategies that employ T-8 lamps. They can be used to uniformly reduce power in overlit spaces without the light and dark patterns caused by delamping. A modern office can be lit using 1 W/ft^2 of power or less. Existing lighting systems in typical office buildings may be using 1.5 to 2 W/ft^2 when 0.85 to 1.0 W/ft^2 would be more than sufficient. Lighting energy usage can be further reduced using occupancy sensors and daylighting controls.

T-5 fluorescent fixtures are a good alternative for warehouses. T-5 and T-8 lighting with reflectors and occupancy sensors is being used to replace high-intensity discharge lighting in warehouses. These luminaries turn off when no one is present, resulting in a significant reduction in electricity use.

Pulse-start metal halide fixtures are used in garages. These can be combined with occupancy sensors to idle at 50% power when no one is present. For exterior use, most building and site lighting is metal halide or high-pressure sodium. There are fluorescent fixtures available that will operate down to 0°F. Fluorescent luminaires have better color rendition and do not take 5 to 10 min to restrike after a power failure. Exterior lighting should be controlled with photocells or astronomic time clocks.

Plug Loads

These items—anything you plug into an electrical outlet—use electricity directly and also impact the loads on the heating and cooling systems by producing heat. Most plug loads are connected by the tenants and are not controlled by the building owner. Additionally, most building owners and managers do not realize the size of what are called *phantom electrical loads* in buildings. Those loads are caused by some electrical devices that consume power when turned off to provide the convenience of what is called *instant on*. Other items, such as chargers on

Figure 3-4. Different lighting strategies utilizing T-8 lamps are used in each of the four office spaces.
(Photograph courtesy of Philips Day-Brite.)

emergency lighting and dry transformers located in building electrical rooms, are continually drawing power.

The miscellaneous equipment within a building can consume considerable amounts of energy. It is added as needed and is so varied that it is difficult to manage as an energy-using category. However, individual pieces of equipment can have improved energy efficiency levels. ENERGY STAR-qualified products can use as much as 60% less electricity than standard equipment. Many landlords now require the installation of only ENERGY STAR-qualified appliances. Organizations such as the American Council for an Energy-Efficient Economy and the Green Electronic Council's EPEAT provide resources for identifying the most energy-efficient equipment. Office equipment such as computers, copiers, and printers are rated, as well as appliances such as refrigerators and coffee machines. To some extent, the loads from these appliances can be managed. Computers' energy can be reduced using models that have low power draw when not in use. Some equipment can be shut down completely when not in use to avoid the small power draws often associated with power supplies and standby modes. Energy use for equipment generally follows the occupancy schedule of the building, with a portion of the load occurring continuously. For example, computers and copiers should be ordered with "sleep mode" features to reduce power consumption when they are not in use. Similarly, vending misers reduce the operation of lighting and compressors in vending machines. Both "sleep mode" features and vending misers have very favorable returns on investment. For equipment without sleep modes, there are power strips fitted with occupancy sensors that will shut equipment off automatically when no one is present.

Plug loads are a category of energy use that is growing. There is not a lot of consumption data available for plug loads. Plug loads typically account for about 0.5 to 2.0 W/ft^2 in an office building. They can be measured individually. Inexpensive data loggers are available that plug into a wall outlet. The plug load is plugged into the logger and the power use is recorded. Alternatively, measurements can be made at the circuit breaker panel, provided there are dedicated circuits for the plug loads.

Building Envelope

The building envelope does not directly use energy but is a passive element in defining the loads placed on energy-using systems within the building. The thermal performance of the roof, walls, and windows drive some of the heating and cooling requirements. Their importance is emphasized in smaller buildings where the envelope area of the building is larger when compared with the building's volume. Energy use related to the building structure is mostly correlated with outdoor temperature, resulting in larger energy use during weather extremes.

The building envelope components can be thought to have efficiency levels rather than system impacts. Upgrades are typically capital intensive and are made as part of remodeling efforts (e.g., added roof insulation, window replacement). There are some augmentations that can be applied to improve envelope performance. Window films or shading devices can reduce the heat gained from the sun. A couple of shading options are shown in Figure 3-5. Air sealing and addressing heat losses/gains around penetrations can be performed as maintenance items. For the most part, the efficiency of the building structure is designed into the facility and is not easily changed.

The structure serves as the barrier to the outdoor environment, shielding the interior from outside elements such as moisture, wind, dirt, and noise, and providing a thermal barrier keeping conditioned air in the building. When making modifications, one must consider the impact on all of the functions of the structure along with the impact on thermal performance. For example, changes might impact moisture/vapor control and transfer, causing unwanted condensation on the surface or within the structure.

Distribution of Heating and Cooling

The ability to heat and cool must be combined with a means to move heating and cooling to where it is needed. The distribution system also serves to dilute pollutants and to provide homogeneous comfort levels, usually while attempting to minimize noise levels. The distribution system consists of fans and pumps and

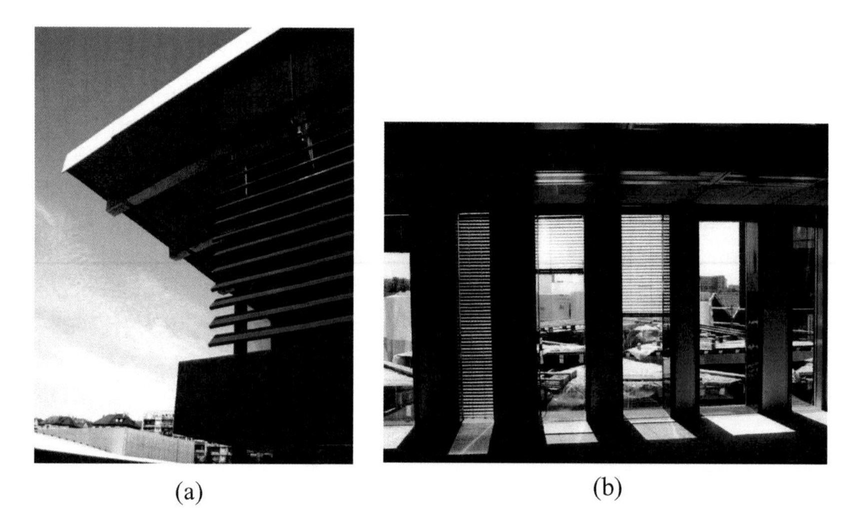

(a) (b)

Figure 3-5. (a) External shading option controlled based on solar position and intensity, and (b) motorized blinds that automatically control incoming light.

(Photographs courtesy of *High Performing Buildings* [Parsley and Serra 2009; Hovorka 2008].)

associated ductwork and pipes. These pieces of equipment have their own inherent efficiencies along with the motor efficiency that drives them. Motors can be upgraded to premium efficiency levels and fans and pumps can be modified to more closely match their loading with their most efficient operating point.

Distribution systems should be reviewed to determine whether

- pumps and fans have premium efficiency motors and are fitted with variable frequency drives (if not, can constant flow systems be converted to variable flow?),
- air-handling systems have economizer cycles to provide free cooling when outside conditions permit (economizers and sensors should be operational; the damper linkages are to be connected to the damper motor and not restricted from movement),
- the air distribution system is balanced properly (if not, are some spaces over conditioned (heated or cooled) to make up for deficiencies in other areas of the building?),
- the existing/available controls are being employed most effectively, and
- air-to-air energy recovery can be added to reduce heating and cooling loads attributable to outside air.

The operating conditions and scheduling of distribution systems can play a large part in their energy use. The ability of the distribution system to operate only when heating and cooling is needed and to modulate with the level of conditioning required will minimize energy use. Systems that throttle flows, rather than reduce the speed of the drive, will consume more energy. Some systems are throttled severely due to changes in loads from those of the original design. These are candidates for modifications to fans and pumps to more closely align their operating points with peak efficiency levels. Efficient operation is achieved through the use of controls and variable-frequency drives. In some instances, multiple modular systems such as modular boilers will operate more efficiently than one large system.

Options including heat wheels, heat pipes, and runaround loops can often be incorporated within the distribution system to provide free-cooling from ambient conditions (commonly referred to as an *economizer cycle*). These systems bring in larger amounts of outdoor air when the cooling requirements can be met with cooler outdoor air. Their operation must be periodically reviewed, as failures will cause the cooling system to operate as normal but use more energy than necessary. Energy use for distribution systems follows the building occupancy schedule with some variation due to modulation with heating and cooling loads. These systems typically operate to meet occupant needs, so their scheduling and control points should be reviewed to avoid unnecessary and wasteful operation.

Ventilation Air. The introduction of outdoor air is important to occupants' well-being. Minimum levels are defined in codes and standards (i.e., *ANSI/ASHRAE Standard 62.1-2007, Ventilation for Acceptable Indoor Air Quality* [ASHRAE 2007c]) and have varied widely over past decades. Outdoor air can have a big impact on energy use, as the air must be conditioned to be acceptable to the indoor environment. Energy use for outdoor air treatment is proportional to weather conditions. Systems can be controlled to respond to temperature or to both temperature and humidity levels. Care must be taken to minimize latent load increases from outdoor air in warm humid climates. When energy prices escalate, it is tempting to limit the amount of outdoor air introduced into the building. Yet this can lead to health and comfort issues for the occupants. Pretreating outdoor air with energy recovered from exhaust streams is often considered in new construction and equipment replacement. Systems that do this are called *energy recovery systems*, but cooling savings often result as well. Types of energy recovery systems include heat recovery wheels, heat pipes, cross-flow heat exchangers, and runaround loops (see Figure 3-6). These systems can increase the energy efficiency of treating outdoor air by 50% to 75%, along with an associated small increase in fan or pump energy for moving air through the equipment.

Incorporation of energy recovery into an existing system may be considered when equipment is being replaced. The prerequisite for energy recovery is that an exhaust airstream is close in proximity to the intake airstream, although there are systems that can pump a fluid between separate intakes and outlets.

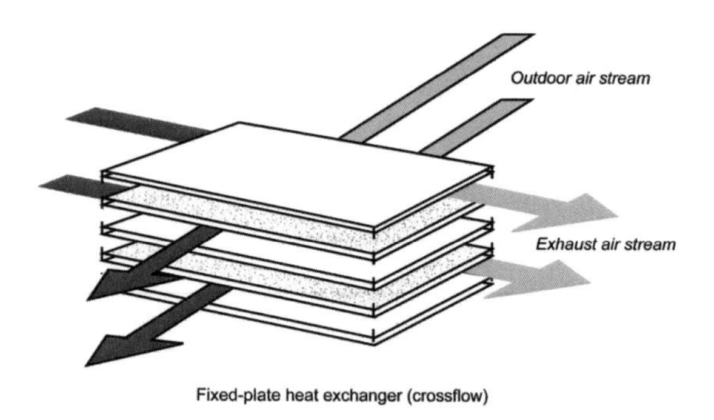

Fixed-plate heat exchanger (crossflow)

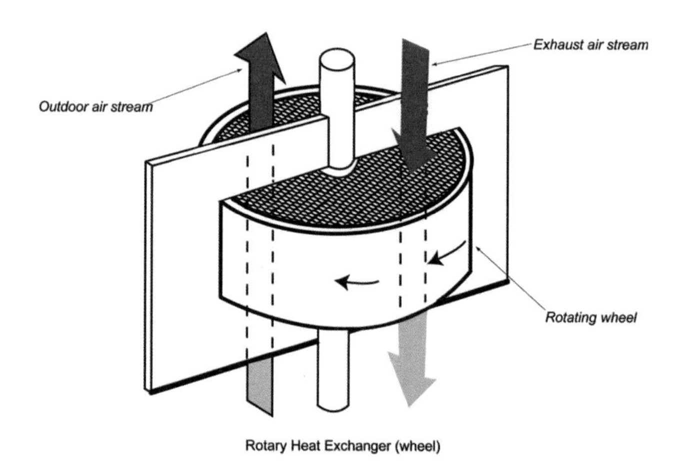

Rotary Heat Exchanger (wheel)

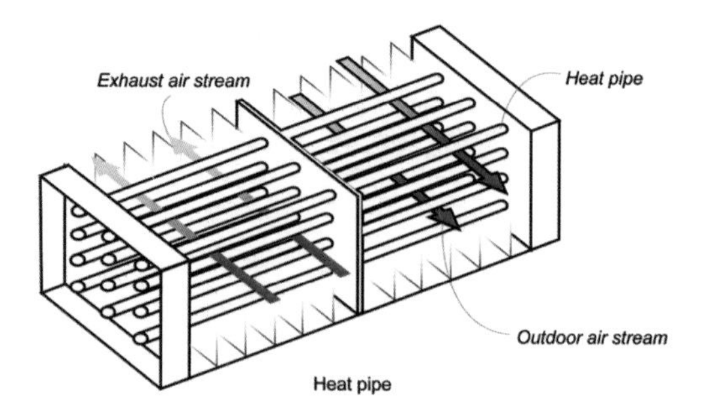

Heat pipe

Figure 3-6. Types of energy recovery systems.

In some applications with widely varying occupancy, the outdoor air can be controlled based on the number of people in the space. This application, known as *demand-controlled ventilation*, employs carbon dioxide (CO_2) monitoring to limit energy use during times of minimal occupancy. For example, when a building is unoccupied, the outside air damper is closed. When the building is occupied, outside air is delivered in the amount required to maintain CO_2 at safe levels for the actual occupants, not the minimal amount of air for the design maximum occupancy. This method reduces the amount of air that would be delivered and conditioned (based on design occupancy) by delivering the right amount of air for the actual occupancy.

Water Heating

Temperature and quantity are the two major concerns when it comes to service or domestic water heating. Water heating is often provided through standalone equipment but can be provided by heating equipment primarily used for space heating. Systems often incorporate storage volumes to accommodate spikes in hot water draws and continuous circulation to accommodate instant hot water at faucets. Some systems distribute instantaneous units throughout a facility to directly meet loads without losses from storage or circulation. Losses decrease with decreases in storage temperature, but a sufficiently high temperature is required to prevent the growth of biological contaminants. Energy use for water heating is highly dependent upon the application. Offices might see a fairly consistent pattern (albeit low consumption level) of energy use during occupied periods, while residential facilities see energy use correlated to water use for bathing and kitchen activities. Within commercial buildings, specialized end uses can be drivers in larger water heating loads and may be best served by dedicated service hot water systems. Such buildings include restaurants, food service operations, and laundries.

Water-heating distribution systems with circulation pumps should be inspected to ensure that the piping is insulated and the pump is controlled by a means such as a reverse-acting return line aquastat. This device provides better control and more energy savings than a time clock. Such units function by turning the pump off when there is no demand for hot water in the building.

Sometimes it makes economic sense to replace a central domestic hot water system with a point-of-use system. Many facility types have extremely long lengths of domestic hot water supply and return piping. In large buildings, this can run from thousands of feet to miles of pipe. Much of this pipe is buried in walls and is, hence, practically inaccessible. In older buildings, it is a good bet that this piping is not insulated. Given the small use of domestic hot water in office buildings, the question arises as to whether it is more cost effective to run 125°F or hotter water through all of that piping or install point-of-use water heaters.

While water is not energy, water use is another utility to the building owner. Water conservation through the use of low-flow toilets, low-flow or waterless urinals, low-flow faucets and showerheads, graywater systems, and reduced irrigation strategies can be cost-effective means of reducing the operating costs of a building. (See Figure 3-7.)

Refrigeration

Refrigeration systems are limited to specialized applications (e.g., supermarkets, food processing, and coolers). As with cooling, there are many choices when it comes to compressors and refrigerants, and there are system options that help to recover waste heat and maximize the use of cooler outdoor conditions. These systems must also defrost to eliminate the buildup of ice that insulates the cooling coils within the display case or cooler. Refrigeration systems can respond to refrigeration loads with maximum efficiency by staging multiple pieces of equipment or by using equipment driven at varying speeds. System efficiency is also dependent on the temperature of the coolers. Operating equipment at the highest temperature to meet the load optimizes energy use. In some instances, heat can be recovered from the compressors and used elsewhere in the building.

In supermarkets, energy can be saved by reducing the use of heat within the refrigeration case. If existing cases have lighting and antisweat heaters, these can be turned off by timer controls when the premises are closed. Alternatively, triple glazing can be installed in lieu of antisweat heaters in the glass

(a) (b)

Figure 3-7. (a) Waterless urinals and (b) motion-activated, ultra low-flow faucets help conserve water.

(Photographs courtesy of *High Performing Buildings* [Schreiber 2009].)

doors of refrigerated display cases. The heat caused by display lighting in refrigeration cases increases the amount of refrigeration energy required. To address this problem, LED lighting can be used instead of fluorescent lighting, since it produces less heat. With open display cases, the space is overcooled by the display case, and HVAC systems must be designed to provide dehumidification with limited cooling. One way to save energy with open display cases is to cover the cases when the facility is closed.

Heating and Cooling Equipment

Heating and cooling equipment in commercial buildings consists of three parts: prime movers, distribution systems, and controls. The prime movers are the air conditioners, heat pumps, boilers, and furnaces that produce steam, hot or chilled water, or air that is used to condition the building. The distribution system consists of the piping, pumps, ducts and fans, and related valves and dampers that deliver the conditioned water or air to the occupied space. The controls sense the conditions in the building and operate the prime movers and the distribution systems to attempt to maintain desired conditions in the space. All three parts must work in harmony to produce an efficient heating and cooling system.

The simplest cooling system is a window air conditioner. A window air conditioner cools refrigerant, a fan blows air over the refrigerant to cool the space, and the waste heat is rejected outside. For medium-size systems, variable refrigerant flow is gaining popularity. Variable refrigerant systems link multiple room units into a combined system, which is more efficient than stand-alone units.

With large commercial systems, it isn't practical to circulate refrigerant all over the building. Chilled-water systems, as shown in Figure 3-8, are used instead. A chiller produces cold or chilled water that is circulated to coils in the building's air-handling units. The condenser or cooling tower rejects heat, much like the coils on the back of a refrigerator. The chiller still contains a refrigeration cycle like an air conditioner. However, cold refrigerant is used to produce chilled water, which is distributed to the cooling coils of air-handling units. The chiller produces waste heat in the process of generating the chilled water. The waste heat can be removed directly by blowing air over condenser coils located outside of the building. In large systems such as the one represented in Figure 3-8, the waste heat warms condenser water, which is cooled in a cooling tower.

Rooftop air-cooled cooling equipment, as shown in Figure 3-9, is simpler than a water-cooled chilled-water system but is not as energy efficient.

Equipment Efficiency Rating. The inherent efficiency of heating and cooling equipment is often cited when looking for energy savings. Minimum efficiency levels for new equipment are mandated in codes and legislation. Industry and rating organizations pay careful attention so as to define precise conditions under which efficiency is measured so equipment energy use can be compared on an

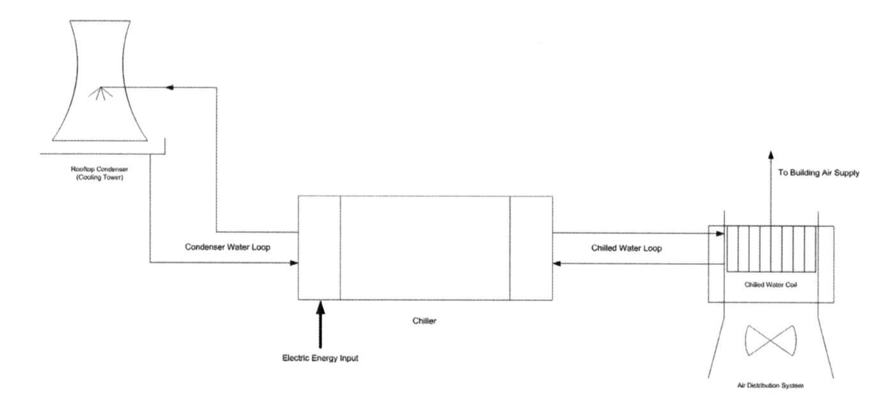

Figure 3-8. Typical chilled-water system.
(Figure courtesy of Fred Goldner, Energy Management and Research Associates.)

Figure 3-9. Rooftop air-cooled cooling equipment.
(Photograph courtesy of Fred Goldner, Energy Management and Research Associates.)

equal basis. These efficiencies often refer to the operation at design or peak conditions, but equipment must often operate over a wide range of conditions.

Equipment Operating Efficiency. Too often in the design phase, focus is placed on ensuring that the building systems perform well at full load. The heating and cooling equipment is designed to maintain comfort on the hottest and coldest days of the year and to supply the heating and cooling capacity required to recover from night setback conditions. However, the hottest and coldest temperatures occur for very few hours of the year. Most of the year is spent operating at a fraction of the system peak loads. Therefore, part-load operating efficiency is more important than peak-load efficiency with regard to building heating and cooling systems. Inefficiencies often occur in the transitional seasons when both heating and cooling may be required and when systems are operating at low partial loads. The operating conditions and the ability of the equipment to adjust to those conditions are significant factors in determining energy use. To provide efficient operation over the full range of conditions that are encountered requires either built-in capability to modulate with load (e.g., variable-frequency drives on motors and variable-air-volume on fans) or multiple pieces of equipment that are properly staged "on" or "off" as loads vary.

Take, for example, a space heating boiler that is 90% efficient. That efficiency occurs at steady-state operation, meaning the boiler is constantly firing. That situation occurs on a very cold day. On warmer days, the boiler fires at a lower rate. This is similar to turning down the flame on a gas stove. The operating efficiency is still pretty high. Eventually, it is so warm that the boiler must cycle on and off. This results in a drop in operating efficiency. When a heating boiler is fired, say only a few times per day to make domestic hot water in the summer, the efficiency is quite low—maybe 30%. So, over the course of a year's worth of providing heating and domestic hot water, the combustion efficiency of the 90% efficient boiler may be only 70% or 75% or less. The seasonal efficiency of a hot water boiler system can be increased by using condensing boilers (i.e., a boiler that recovers energy that is discharged through the flue of a conventional boiler) that can operate at low part loads or by installing multiple modular boilers.

Equipment Technology. The choice of technology can also play a large role in the energy use of the equipment. For example, cooling systems have many options, each with positive and negative attributes regarding first cost, energy use, maintenance requirements, precision of control, and ease of operation. For existing buildings, replacement of equipment within the same family of technology is often a priority unless a major renovation or remodel is undertaken. Improvement opportunities are often overlooked when a simplistic "replace with like" approach is taken. Instead, the operating cost over the life of the piece of equipment should be considered.

Consider, for example, installation of an electric chiller vs. a gas-fired absorption chiller. The electric chiller is cheaper, requires less maintenance, and requires a smaller cooling tower. It is also easier to modulate under part-load conditions. The gas-fired chiller uses a less expensive fuel and greatly reduces peak demand electric charges. Thus, the lowest life-cycle cost alternative is not immediately clear. What about water-cooled vs. air-cooled chillers or steam or double effect absorption chillers? There are many choices in the marketplace, any of which may be a best fit for a particular application. Sometimes a combination of technologies is the most cost-effective choice.

The heating and cooling equipment must deal with space humidity requirements to maintain comfort levels in the building. Most systems cool the air and simply rely on the associated dehumidification, allowing the humidity levels to vary with cooling load and equipment operation. Some applications, such as laboratories, factories, or museums, require a controlled humidity level that requires additional consideration of the cooling/dehumidification equipment. The market offers a range of technologies for dehumidifying (e.g., overcooling and reheating and desiccants that absorb moisture directly from the airstream), each with its own advantages and applications.

Humidification can add moisture to the airstream at a central location or in specific spaces. As with dehumidification, the market offers a range of technologies (e.g., heating water from electricity or gas and atomization using pressure and nozzles) for humidification applications.

There are a variety of alternative energy and combined heat and power systems that can be used to reduce energy costs. Geothermal (ground-source) heat pumps are a mature technology and are very energy efficient. They have a higher first cost than traditional HVAC systems; however, the earth loops last 50 years or more, and the requirements for mechanical space in the building are smaller than those for boiler and chiller systems. Exterior space is required around the building for installing the earth loop, and low-rise applications tend to be more cost effective.

Photovoltaic (PV) systems can be used to add a percentage of green power to a building. The prices of these systems are decreasing, and when combined with tax credits, PV can be a cost-effective alternative. But, it is important to determine that the local solar resource is sufficient to provide cost-effective power. (See Figure 3-10.)

Cogeneration systems such as natural gas engine generators, microturbines, and fuel cells generate electricity and produce waste heat. These systems can be cost-effective if there is a sufficient cost differential between natural gas and electricity and if there is a year-round need for the waste heat in the building.

REDUCING ENERGY COSTS

It is possible to reduce energy costs without reducing energy use. There are three ways to accomplish this: shop the commodity, reduce the peak demand charge, and shed load when asked by the utility.

Figure 3-10. A roof-mounted 20 kW PV system provides power back to the grid.
(Photograph courtesy of *High Performing Buildings* [Anonymous 2008].)

1. **Shop the Commodity:** Energy is deregulated in many jurisdictions and savings can be achieved by procuring energy from a supplier other than the distributing local utility. In the case of electricity and natural gas, the commodity will still be delivered by the local utility but may be purchased from a variety of sources.
2. **Reduce the Peak Demand Charge:** Commercial electricity (and sometimes steam or even gas) tariffs are typically divided into consumption charges and demand charges. The demand charge is based on the peak 15 or 30 min usage of the facility in the billing period (usually 30 or 60 days). The rationale for this charge is that the utility companies must build and maintain generation capacity to meet and exceed the highest demand they experience, even if it is only for one hour per year. Money can be saved by controlling when a building uses energy, because most commercial buildings are charged for the peak rate of electricity use in a billing period. Under some rate structures, demand can account for a very significant portion of the bill. Additionally, many electric utilities are switching to real-time pricing—the price of electricity changes every hour. With this change, it will be even more important to be able to control when a building uses energy.

3. **Shed Load When Asked by the Utility:** In many jurisdictions, it is possible to get lower electrical rates or to get paid-for electric capacity that can be turned off if there is a capacity crisis that might result in brownouts or outages. In some cases, the system operator will pay the building owner for joining the program and agreeing to shed load when asked. Additional money is earned if you are called upon to help. This will require turning off nonessential equipment or running emergency generators when asked to reduce the building electric load. The number of hours per year that a building must shed load to participate in the program is typically small, and ample notice is given to permit the building operator to take action. In some years, a building may never be called upon to shed load yet still receives payments. Many electric system operators, such as the New York Independent System Operator (NYSIO), run annual auctions for businesses willing to shed load. The minimum bid is one megawatt, so most buildings work through an aggregator (i.e., a firm that groups and organizes the load shedding of multiple buildings). Some aggregators work with the building owner to identify and quantify sheddable loads that will not interfere with building operations. If a bid is selected, the NYSIO makes a monthly payment per megawatt bid. If a load shortage occurs, a call to shed load is given the day prior to the event. The building(s) can decide how many kilowatts will be shed but must shed the agreed-upon kilowatts of electricity if called upon to do so. Failure to shed load can result in penalties. Events last up to four hours, and payments are made for four hours at the marginal kilowatt per hour cost at the time of the event. This might be $0.50 or $1.00/kWh. If a building participates through an aggregator, that firm will keep at least 10% of the income for coordinating activities. Some aggregators offer optional plans. For example, they may keep a greater percentage of the payments in return for allowing the building to avoid partial or complete shedding if an event occurs at an inconvenient time.

INDUSTRY STANDARDS

Building codes and conditions are assessed against established standards such as *ANSI/ASHRAE Standard 55-2004, Thermal Environmental Conditions for Human Occupancy* (ASHRAE 2004a), ASHRAE Standard 62.1 (ASHRAE 2007c), and the *Illuminating Engineering Society of North America Lighting Design Guide* (2000). These provide reference points for recommended illumination levels of interior spaces. Meeting or exceeding these operational standards should be the first step—and a requirement—in maintaining and improving the operating efficiency of a building. Note that while state and other local codes are usually built upon these standards, there may be parts of local codes that exceed these standards. In such cases, the local codes should be followed.

Making the Business Case 4

INTRODUCTION

For the building owner, investing in building energy efficiency competes with other financial investments. An assessment is necessary to prioritize energy efficiency improvements with other investment opportunities. Additionally, when making building or system upgrades, it is paramount to take an integrated approach and consider the interplay between systems. Business owners and managers should thoroughly evaluate energy-efficient technologies and renewable energy technologies that are not yet the status quo. Doing so will reduce the risk of obsolescence.

When making energy efficiency decisions, building owners and managers face many barriers. To overcome these barriers, they must make a business case for improving the energy efficiency of existing buildings. Making the business case involves evaluating the risks, conducting financial analysis, recognizing the benefits, learning about the incentives, and prioritizing energy efficiency projects as part of an overall capital plan. Savings realized from implementing no- and low-cost operational adjustments and an energy awareness program can be used to fund the cost of an energy audit and more capital-intensive projects.

BARRIERS AND SOLUTIONS TO ENERGY EFFICIENCY

Often-cited barriers to making energy-efficiency upgrades are the following:

- *Capital Constraints:* Available internal dollars have to compete with other allocations and property improvements. Further, the capital dollars earmarked for the property are usually first spent on building improvements that drive revenues such as updating common areas, tenant improvement allowances, and leasing commissions. As the price of energy increases and tenant demand for energy-efficient buildings increases, so will the allocation of capital dollars to energy efficiency. Increased transparency of the energy efficiency of buildings will be the greatest driver of improved energy efficiency in buildings.

- *Collateral Issues*: Most investment property is owned by unrated limited-liability companies with no personal guarantors. Traditional lenders want strong, investment-grade borrowers on equipment loans. This is why government buildings see energy upgrades more readily than the private sector.
- *Holding Period Bias*: The investment property sector is staunchly opposed to encumbering an asset in any way that would limit its ability to be sold quickly. Most holding periods are three to seven years. As a result, the payback period calculation has most often been used for capital improvement analysis. This limits most energy-efficiency upgrades to lighting systems.
- *Split Incentive:* There is no doubt that energy efficiency adds value to real estate. Reducing operating costs increases net operating income. What can cause problems is the issue of who pays and who benefits. Landlords of existing multitenant office buildings have been reluctant to make capital investments in energy- and water-saving technologies because tenants pay the utility bills. Conversely, tenants with triple net leases have been reluctant to improve a landlord's property, even if they are paying the utility bills. Interestingly, most leases allow for the amortization of capital expenditures that improve efficiency and the pass-through of such expenses to tenants in the annual operating expenses. Increased holding periods would provide the landlords with more time to recoup investments through pass-through operating expenses. Individual lease language should be reviewed.

Energy saving performance contracting (ESPC) and the new Building Owners and Managers Association (BOMA) International Energy Performance Contracting Model, as well as the advent of green leases, are providing solutions to these barriers.

Energy Saving Performance Contracting

In a carbon- and capital-constrained market, new mechanisms and stimuli are emerging. Consider ESPC, which is an agreement between a building owner or occupant and a performance contractor. The contractor, referred to as an *energy service company* (ESCO), typically identifies, designs, and installs the EEMs and guarantees their performance or energy savings. In essence, the savings in the utility line item resulting from the EEMs are used to finance the projects. The ESCO, the building owner, or a third party can provide financing. While many public entities have been successfully engaging in ESPC for more than 20 years, the private sector has been very slow to adopt this solution to paying for energy-efficiency upgrades. This lack of adoption stems from complexity of agreements and verification of energy savings, long project timelines, and lenders requiring a lien and/or personal guarantees.

Hiring an Energy Service Company. An ESCO will perform an audit, obtain financing, install improvements, and get paid a contracted monthly amount over a period of time to cover the cost of these services. The payment is, in concept, less than the monthly energy savings, so this is sometimes referred to as a *shared savings program*.

Using an ESCO has advantages and disadvantages. On the plus side:

- The owner gets the benefit of the improvement immediately, in the form of new building systems that operate more efficiently and provide higher comfort levels.
- There is no up-front cost to the building owner.

On the minus side:

- The ESCO is in business to make a profit. It will borrow the money to install the conservation measures and make a profit on the loan.
- Its profit reduces the net return on investment to the building owner and therefore reduces the number of cost-effective measures that can be undertaken.
- The owner will have to sign a contract in which the owner agrees to repay the ESCO on a monthly basis. The owner can agree on the energy savings and lock in a set payment, or a third party or the ESCO can perform periodic monitoring and verification and adjust the payments depending upon how the conservation measures perform. This can become complicated, as the ESCO will not be responsible for the negative impacts of changing building usage patterns.

If the decision is made to use an ESCO, the National Association of Energy Service Companies (NAESCO) can provide a list of reputable ESCOs in the building's local area. It is important to check their references. The owner will be required to sign a contract. An attorney can translate the legalese, but the assistance of an energy consultant may be needed in regard to the measurement and verification (M&V) plan and conservation measure selection. The owner can issue a request for proposal and have multiple ESCOs respond.

Project goals should be defined in terms of

- necessary performance improvements and
- energy and O&M-related cost savings.

M&V requirements need to be established in relation to the potential risk and as a percentage of the project costs. M&V should include

- verification of total energy and cost savings and
- equipment performance (commissioning is desirable for complex measures).

Long-term verification may be desirable for capital-intensive measures.

The BOMA Model. The recently released Building Energy Retrofit Program (BOMA 2006b) produced by BOMA International and the Clinton Climate Initiative provides an ESPC model that offers building owners and operators a simplified, turnkey method for undertaking energy-efficiency projects. According to BOMA International, "in this model the building owner sets the financial and environmental parameters and the ESCO audits the building(s), designs the project, establishes a maximum project cost, calculates minimum energy savings, and acts as the prime contractor for project implementation—providing ongoing monitoring and verification of project performance after completion" (pp. 5–6). This standardized contracting model has been vetted by real estate industry professionals, and the ESCO guarantees the financial savings, compensating the owner for any saving shortcomings and assuming project risk. A copy of the program and all supporting documents can be downloaded free of charge at www.boma.org/RESOURCES/BEPC/Pages/default.aspx.

Green Leases

Green leases are standard leases modified to remove barriers to improved energy efficiency and sustainability. The structure of a green lease is one that motivates all of the parties involved to invest, operate, and work in a building in the most energy-efficient and environmentally responsible manner. Green leases add transparency and accountability to a building's energy performance and environmental impact and the roles played by both landlords and tenants. Green leases are often standard lease forms with riders that communicate shared environmental standards, goals, and responsibilities.

EVALUATING THE RISKS

The risk is not in making an investment in energy efficiency but in failing to become more energy efficient. When evaluating capital investment in energy efficiency, there are at least three major types of risk to consider—market, regulatory, and environmental.

Market Risk

The market is beginning to recognize the importance of and to place value on a building's energy performance level. This is reflected in more stringent energy codes, Commercial Real Estate Information Group reporting of ENERGY STAR-labeled buildings, tenant representation brokers asking landlords to annually disclose a building's energy performance (i.e., its ENERGY STAR score), and states such as California and the District of Columbia mandating such disclosure. Increased transparency and tenant demand will serve as the greatest stimuli for market transformation toward more energy-efficient buildings. Additionally, the cost to certify an existing building under the U.S. Green Building Council's (USGBC)

Leadership in Energy and Environmental Design (LEED) Green Building Rating System can be recovered, in part, through savings in the utility line item. As buildings become more energy efficient, inefficient buildings will become less desirable assets over time.

Regulatory Risk

Buildings consume approximately 70% of the nation's electricity, and much of that electricity comes from fossil-fuel-burning power plants. As we move toward global regulation of energy and carbon in a climate-constrained world, the United States must reduce its dependence on fossil-fuel-burning power plants. As power companies are forced to invest in renewable-energy technologies and carbon becomes a regulated emission, consumers could see pass-through costs. Reducing energy consumption provides immediate savings and mitigates the risk associated with higher energy costs. Further, those owning inefficient buildings may be forced to pay higher rates due to above-normal consumption.

The city of Seattle was the first city to require all city buildings to achieve the USGBC LEED silver rating or higher. The new Seattle City Hall achieve a LEED-NC gold rating.

(Photograph courtesy of *High Performing Buildings* [Bischak and James 2008].)

Environmental Risk

Climate risk is directly impacting insurers. Losses affect the bottom line of insurance companies and, thus, insurance premiums. Insurance regulators and reinsurers (such as Swiss Re) are becoming increasingly concerned that global climate change could potentially lead to a crisis of insurance affordability and availability as well as solvency for the insurers themselves. They are prompting insurers to incorporate climate risk into their underwriting and provide financial incentives to encourage risk-reducing behavior in the form of energy-efficient, green buildings.

CONDUCTING FINANCIAL ANALYSIS

The most common methodologies for analyzing capital expenditures are simple payback period, net present value (NPV), internal rate of return (IRR), and life-cycle costing (LCC). This section briefly defines these methods, discusses the pros and cons of each, and discusses which methods are best suited for analysis of energy efficiency measures (EEMs). The advantages and disadvantages of these methodologies are summarized in Table 4-1.

Simple Payback Period

The simple payback period calculation provides a rough estimate of the time needed to recover the initial investment. It is the total or incremental cost of the project divided by the reduction in energy and other operating costs resulting from the improvement. Total cost is generally used for evaluating EEMs. Incremental cost is used when replacing equipment at the end of its useful life to evaluate higher-cost, higher-efficiency alternatives. For example, assuming the total cost of a chiller replacement is $900,000 and the annual change in cash flow resulting from the new system is $103,680, the simple payback period for the investment is 8.68 years:

$$\text{Simple Payback Period} = \text{Total Project Cost/Annual Change in Cash Flow}$$

$$\text{Simple Payback Period} = \$900{,}000 / \$103{,}680 = 8.68 \text{ years}$$

While simple payback period analysis is the quickest way to evaluate a capital improvement expenditure, it sells short the impact the improvement has on the building's cash flow. Simple payback period analysis does not take into consideration a) the impacts on cash flow beyond the payback period or b) the time value of money. In other words, it does not provide the value of the improvement as reflected by its impact on cash flow.

Net Present Value (NPV)

NPV analysis is the preferred method for evaluating *independent* capital improvement projects. NPV analysis involves taking the cost of the project today (i.e., negative cash flow) and adding to it the present value of the changes in annual cash flow that result from the improvement. Like the simple payback period analysis, NPV calculations require the total cost of the project and the annual changes in cash flow that result from the capital improvement. The difference is that NPV analysis takes into consideration the capital improvement project's impact on cash flow over the useful life of the improvement and the time value of money (Pinches 1984).

Table 4-1. Examples of Financial Analysis Methodologies

Methodology	Advantage	Disadvantage
Simple Payback Period	Easiest to use. OK for paybacks of three years or fewer and minimal maintenance items.	Doesn't consider cash flow beyond the payback period and doesn't include the time value of money.
Net Present Value	Accounts for the time value of money. Use to evaluate single alternatives.	More complex calculation than simple payback period.
Internal Rate of Return	Provides minimum rate required for positive cash flow.	Renders incorrect IRR with nonsimple cash flows.
Life-Cycle Costing	Includes total cost over life of EEM. Use for mutually exclusive alternatives.	Most comprehensive of the methods—useful for evaluating multiple competing EEMs.

The discount rate used should reflect, as a starting point, the organization's cost of capital or the real estate investment's rate of return. The discount rate can be further adjusted to compensate for any perceived risk.

Looking at the same chiller replacement as in the simple payback period example, the present value of future cash flows, over the 20-year useful life of the chiller replacement, is $1,098,387. With the cost of the new chiller at $900,000, the NPV is $198,387:

$$\text{NPV} = \sum_{t=1}^{n} \frac{CF_t}{(1+k)^t} - CF_0$$

$$\text{NPV} = \sum_{t=1}^{20} \frac{\$103,680_t}{(1+0.07)^t} - \$900,000$$

$$\text{NPV} = \$198,387$$

where

CF_t = changes in annual cash flow
CF_0 = initial change in cash flow (i.e., total cost of investment)
k = the discount or hurdle rate
n = the number of years of useful life of the improvement
t = time period

Interestingly, most projects would have been forgone with a payback of 8.68 years, yet this capital project adds a value of $198,387.

The higher the NPV, the greater the profitability of the investment. The decision rules for NPV analysis are as follows:

- If the NPV is greater than zero, then accept.
- If the NPV is less than zero, then reject.
- If the NPV is equal to zero, then accept or reject based on factors other than economics.

Internal Rate of Return (IRR)

The IRR is the rate of return on the investment that yields an NPV of zero. The use of IRR analysis is not ideal in making capital budgeting decisions. However, IRR analysis can be used to supplement an NPV calculation; it should not be used to replace NPV. The preference of NPV analysis over IRR analysis is due to problems that arise when a) there are nonsimple cash flow series (positive and negative cash flows) and/or b) the analysis is being used to rank mutually exclusive projects, which is often the case when evaluating and choosing EEMs (Pinches 1984).

Life-Cycle Costing (LCC)

Once an EEM that requires capital dollars has been identified, the next step is to evaluate alternatives to determine the best investment. LCC analysis is best suited for comparing alternative choices for a single EEM. The analysis requires calculating the present value of all costs associated with the project over its life cycle—acquisition, installation, removal of existing equipment, soft costs (e.g., consulting fees, energy modeling, design fees, etc.), cost of energy and energy consumed, maintenance, financing, incentives, and disposal. The alternative with the lowest LCC (the lowest cost) is the best investment.

LCC analysis is ideal for evaluating mutually exclusive alternatives for a single EEM, such as installing equipment that either a) meets minimum code requirements, b) is high efficiency but has a higher first cost, or c) is an on-site renewable energy source. For example, LCC analysis can be used to determine the most economically prudent seasonal energy efficiency ratio (SEER) rating of a heating, ventilating, and air-conditioning (HVAC) unit, or the R-value (the R-value is a measure of the resistance of a material to the transmission of heat measure in $ft^2 \cdot h \cdot °F/Btu$) of roof insulation, or the shading coefficient of a window (the ratio of solar energy transmitted to incident solar energy).

Variations of LCC analysis can also be used to rank or prioritize EEMs based on budget constraints. LCC is the present value of all costs associated with the project. For ease in ranking EEMs, LCC analysis provides greater flexibility and ease of analysis.

For accurate comparisons of EEMs when employing present value techniques, either for NPV or LCC analyses, basic principles of capital budgeting must be employed and include the following:

- Adjustments must be made for mutually exclusive projects with unequal lives. If comparing mutually exclusive projects that have different useful lives, it is necessary to equalize the lives of the alternatives to make a valid comparison. This can be done mathematically or by adjustments to cash flow assumptions.
- Sometimes a project with a large capital cost will have an NPV that is only slightly better than a project with a low capital cost. In that case, the owner may opt for the lower cost option to free up capital for other investments.

LCC analysis should be used in both new construction and existing building upgrades, and it should be integrated early into the design process.

The U.S. General Services Administration (GSA), as well as state and local governments, use LCC for economic analysis. The GSA uses LCC as prepared by the National Institute of Standards and Technology (NIST). NIST published the *Life-Cycle Costing Manual* for the Federal Energy Management Program (Fuller and Petersen 1995) and annually releases *Energy Price Indices and Discount Factors for Life-Cycle Cost Analysis* (Rushing and Lippiatt 2009). NIST also established the Building Life-Cycle Cost computer program to perform LCC calculations for comparing mutually exclusive alternative projects as well as ranking independent capital projects (2009). These resources were used in the development of this section and are available for free download from the FEMP Web site at www1.eere.energy.gov/femp/program/lifecycle.html.

Calculating Cash Flows

Regardless of the financial analysis methodology used, (i.e., simple payback period, NPV, IRR, or LCC), the critical step is to establish accurate assumptions regarding the annual cash flows. Assumptions must be accurate, or results will be garbage. Account for all of the impacts on cash flow including the energy savings, changes in operations and maintenance (O&M) costs, interactions among system changes, incentives, financing costs, and taxes.

Calculating the Energy Savings of an EEM

While most of the variables used in determining cash flow are readily quantifiable—acquisition, installation, disposal, and soft costs—the energy and water savings component and the impact on utility bills are by far the most challenging to quantify, but they are the most important. The value assigned to the energy or water savings component of an EEM is typically provided by the vendor selling the product. This often poses a barrier for the decision maker, as purported savings based on overly optimistic assumptions might be skewed to favor an inordinately expensive piece of equipment.

Calculating energy savings is the most difficult part of analyzing a capital-intensive energy-efficiency project. Calculating energy savings can be exact or rough estimates, depending on the amount of information available. Variables needed to determine the energy savings of an EEM typically include:

- the current and projected cost of energy (both on-peak and off-peak),
- the hours the equipment operates (runtime or operating hours) and when it runs (on-peak and off-peak),
- the efficiency of the equipment and the units of energy it uses (i.e., kW/ton or W/lamp), and
- the indirect efficiency gained by the EEM (such as lighting upgrades, window treatments, and cool roofs that reduce the demand on the building's cooling load).

When vendors provide costs/benefits for EEMs, building owners and managers should be sure that vendors provide the breakdown for the variables that make up the energy savings calculation. Owners and managers should consider the following: What are the sources? Are the variables measured and verified, or are they assumptions? If they are assumptions, then owners and managers should ask for the basis and back-up for the assumptions. When evaluating EEMs, the higher the cost of energy and the greater the hours of operation, the higher the energy savings and the sooner the payback.

Example: Chiller Replacement

In the previous simple payback period financial analysis of whether or not to replace a chiller, the $103,680 used to represent the change in cash flows included only the energy savings between the old and new systems. Below is a simplified calculation used for determining the energy savings of the equipment itself.

Variables:

- annual savings = $\$_S$
- building load of 1200 tons remains the same for both alternatives
- efficiency of existing system (as provided by manufacturer product sheet) = 0.72 kW/ton
- efficiency of new system (as provided by manufacturer product sheet) = 0.40 kW/ton
- current cost of energy based on utility contract = $0.12/kWh
- annual runtime of chiller (i.e., operating hours) = 3000 h/yr
- load factor = 0.75, where load factor is the average load experienced by the chiller over the cooling season divided by the peak capacity of the chiller

The annual energy savings between the new and old systems would be

$$\$_S = \frac{(0.72\ \text{kW/ton} - 0.40\ \text{kW/ton}) \times (1200t) \times (3000\ \text{h})}{\times\ 0.75 \times (\$0.12/\text{kWh})\text{kWh}_s} = \$103{,}680$$

Note that if demand charges are paid on the utility bill, demand cost savings would also occur.

Annual adjustments to cash flow should include projected energy prices and other variable changes such as maintenance costs over the useful life of the equipment. If using LCC analysis, the annual energy costs of each alternative would be calculated. The *Annual Energy Outlook* produced by the U.S. Department of Energy (DOE) provides regional and national price trends and can be downloaded from DOE's Energy Information Administration (EIA) Web site at www.eia.doe.gov (2009a).

Example: Lighting Retrofit

Calculating the energy savings of a lighting retrofit is similar to doing so for a chiller replacement. First, the efficiency (total watts) of the lighting system has to be determined. In the case of lighting, this requires an inventory of the light fixtures and the lamps per fixture for each space in the building. Replacing old fixtures and lamps with new, more efficient fixtures will reduce the wattage used. The vendor should provide a detailed inventory by space of the old and new fixtures and lamps and corresponding wattage for each lighting system, as shown in Table 4-2A and 4-2B. Manufacturer product sheets can be used to verify the total watts per lighting system. Calculating the energy savings or energy consumed requires calculating the total wattage used by the lighting system and dividing it by 1000 to obtain kilowatts. Take the total kilowatts per system times the operating hours for each space type to obtain the kilowatt hours (kWh) per space; KWh times the cost of energy in kWh equals the total cost of energy for the lights. Excess lighting power has the secondary impact of adding heat to the space. Reducing lighting wattage will decrease the building cooling load and increase its heating load.

Energy Calculation Tools

The example above energy savings calculations provided are simplified; others can be very complex. Many utility companies have rebate programs tied to the energy savings component. As a result, they offer robust energy savings calculators to accurately determine the energy savings. California, for example, has free online tools for calculating energy savings and training documents for first-time users. These are available on the California Commissioning Collaborative Web site in the Tools & Templates section at www.cacx.org/resources/rcxtools/spreadsheet_ tools.html.

Table 4-2A. Lighting Survey and Recommended Retrofits

Location	Fixture Quantity	Fixture Description	Watts/ Fixture	Installed Watts	Annual Operating Hours
Pre-Retrofit Conditions					
Storage	2	1I75	75	150	2600
Hallway	1	1I75	75	75	2600
Boiler Room	12	1I150	150	1800	2600
Storage	9	1I75	75	675	2600
Storage	1	1I75	75	75	2600
Men's Room	16	1I75	75	1200	2600
Exits	16	2I20	40	640	8760
Storage	1	2F40T12	96	96	2600
Women's Room	1	1I60	60	60	2600
Stairs	3	1I75	75	225	2600
Post-Retrofit Conditions					
Storage	2	1CF18	18	36	2600
Hallway	1	1CF18	18	18	2600
Boiler Room	12	1CR23	23	276	2600
Storage	9	1CF18	18	162	2600
Storage	1	1CF18	18	18	2600
Men's Room	16	1CF18	18	288	2600
Exits	16	2LED2	4	61	8760
Storage	1	2F32T8E	60	60	2600
Women's Room	1	1CF15	15	15	2600
Stairs	3	1CF18	18	54	2600

The DOE has easy-to-use energy cost calculators for replacing motors, chillers, heat pumps, boilers, appliances, lamps, food service equipment, plumbing, and more. These free online calculators are available at DOE's FEMP Energy-Efficient Products Web page, www1.eere.energy.gov/femp/procurement/eep_eccalculators.html.

Useful Life and Maintenance Cost Estimates

Changing systems will almost always impact maintenance costs. Should assistance be needed in determining or verifying the useful life of equipment or maintenance

Table 4-2B. Summary of Lighting kW and kWh Savings

Location	Lighting kWh Savings Mode			kW Savings	
	All Periods	Heating Mode Only	Cooling Mode Only	Installed Lighting	On-Peak
Storage	296	98	135	0.11	0.00
Hallway	148	47	70	0.06	0.06
Boiler Room	3962	1254	1874	1.52	1.52
Storage	1334	422	631	0.51	0.51
Storage	148	47	70	0.06	0.06
Men's Room	2371	751	1122	0.91	0.91
Exits	5074	1939	2025	0.58	0.58
Storage	94	30	44	0.04	0.04
Women's Room	117	37	55	0.05	0.05
Stairs	445	141	210	0.17	0.17

costs of HVAC equipment, the American Society of Heating, Refrigerating and Air-Conditioning Engineers, Inc. (ASHRAE) maintains a database that provides current information on the service life and maintenance costs of typical HVAC equipment at www.ashrae.org/database.

Interactions between Systems

EEM evaluation should reflect the interplay between systems. For example, whether window film is installed or a window is replaced will impact the building's cooling load. When one project's cash flow stream is significantly dependent on another project's cash flow stream, the projects are said to be interrelated. If one EEM significantly impacts another EEM, then they should be evaluated both independently and together. For example, if EEMs address both roofing and HVAC systems, then evaluate the HVAC system assuming the impact of a cool roof and a traditional roof and vice versa.

Interactions also impact equipment specifications. Changes in the building envelope, for example, can reduce the building's required tonnage for cooling. So, installing a cool roof will not only reduce the cost of HVAC equipment but also its annual energy consumption. Load reduction should be addressed first, then equipment upgrades.

RECOGNIZING THE BENEFITS

Building energy systems, over the course of their lives, use energy and require periodic maintenance. The lowest cost alternative is rarely energy efficient. It often has a shorter service life than more expensive alternatives and can be more expensive to maintain. Use of financial analysis helps to ensure that the energy conservation alternatives that are implemented represent the best return on investment for the building owner.

LEARNING ABOUT THE INCENTIVES

Owners and managers should contact local utility companies to find out about available demand-reduction incentives, rebates, and no- and low-cost assistance programs. State and regional energy efficiency programs and incentives may also exist. The next few paragraphs describe some existing programs, plus additional incentives and programs are expected to become available in the future. The Database of State Incentives for Renewables & Efficiency (North Carolina Solar Center 2009) (www.dsireusa.org) is a comprehensive source of information on local, state, federal, and utility incentives that promote renewable energy and energy efficiency.

Federal legislation as of mid-June 2009 includes the energy-efficient commercial buildings tax deduction. This legislation provides a tax deduction of up to \$1.80/ft^2 for owners or tenants of new or existing commercial buildings that are constructed or reconstructed to generate savings of at least 50% in heating, cooling, ventilation, water heating, and interior lighting energy costs of a building when compared to a base building defined by the *ANSI/ASHRAE/IESNA Standard 90.1, Energy Standard for Buildings Except Low-Rise Residential Buildings* (ASHRAE 2007d). Partial deductions of \$0.60/ft^2 can be taken for improvements to one of three building systems—the building envelope, lighting, or the heating and cooling system. The provision is effective for property placed in service after December 31, 2005, and prior to December 31, 2013. In application, it has been very difficult for owners and managers to meet the 50% savings, except in the case of lighting retrofits.

Title IV of the Energy Independence and Security Act, signed in December 2007, sets forth various green building standards for federal buildings. Section 422 of the act establishes the Net Zero Energy Commercial Buildings Initiative to develop and disseminate technologies, practices, and policies for the development and establishment of net zero energy commercial buildings. For more information, visit DOE's Building Technologies Program Web page (2009a) at www.eere.energy.gov/buildings/about.html.

The Energy Improvement and Extension Act of 2008 (2008), signed in October 2008, extends and enhances tax credits and financing related to renewable energy and energy efficiency. The act provides accelerated depreciation for utilities installing smart meters and smart grid systems and provides production tax credits for renewable energy. On the property side, tax credits are available for energy-efficient and on-site renewable technologies including solar power, wind power, wind turbines, fuel cells, microturbines, certain combined heat and power systems, and geothermal heat pumps. For more information on federal tax breaks, visit the DOE Consumer Energy Tax Incentives Web page (2009b) at www.energy.gov/taxbreaks.htm.

PRIORITIZING PROJECTS AS PART OF A CAPITAL PLAN

After conducting the financial analysis of the EEMs identified by the audit or retrocommissioning, develop an energy improvement plan as part of the building's overall capital improvement plan. The prioritization of projects should be based on return on investment and should reflect the remaining useful life of existing equipment and overall spending constraints. When developing energy improvement capital plans, it is paramount for building owners and managers to remember the interplay between building systems. After making operational adjustments and implementing an energy awareness and procurement program, the next step is to make the capital improvements that impact the building's heating and cooling load such as lighting, building envelope, and plug loads. Then, owners and mangers should invest in improving the efficiency of the building's air distribution system by, for example, adding controls and monitoring devices and upgrading fans and motors. Lastly, capital improvements should be made to the HVAC system. The idea is that once other energy efficiency improvements are made, opportunities for right-sizing the equipment often exist. Equipment oversizing is a significant contributor to excessive energy use. It's important to ensure that vendors size new equipment properly by conducting a thorough analysis of the building load. The U.S. Environmental Protection Agency (EPA) reports that chillers are oversized by 50% to 200% and ventilation systems have upwards of 60% more capacity than what is utilized (White 2007). Right-sizing will save money when purchasing equipment, as well as when it comes to ongoing energy demand.

Owners and managers need to take advantage of equipment failures by not accepting like-for-like replacements. Instead, they should identify and evaluate energy-efficient and on-site renewable energy technologies by using LCC analysis and taking advantage of available incentives.

Developing an Effective O&M Program

5

INTRODUCTION

Often, a poorly designed building with good operations and maintenance (O&M) practices will outperform a well-designed building with poor O&M practices. How a building is operated can save an estimated 5% to 20% on energy bills without a significant capital investment, according to the U.S. Department of Energy (DOE) (2009c). The Building Energy Efficiency Program (BEEP) (2006a), offered by the Building Owners and Managers Association (BOMA) International, cites similar savings.

Having a strong O&M program can result in energy efficiency, energy demand reduction, enhanced equipment performance and reliability, building systems lasting for their maximum life spans, and occupant comfort. All of these impact the bottom line. The key elements of an effective O&M program are the following:

1. Elevating the importance of energy management within the organization by appointing an energy manager
2. Requiring a systems manual
3. Focusing on efficient operations in O&M
4. Investing in training
5. Insisting on performance tracking and reporting

ELEVATING THE IMPORTANCE OF ENERGY MANAGEMENT

Let's face it: when reading the words "operations and maintenance" it is easy to get distracted by something more interesting. That being said, it's the "O" in O&M that is the low-hanging fruit. Therefore, reaching the energy savings embedded within building operations requires energy efficiency to become a priority, without sacrificing occupant comfort. For the O&M department to receive the direction and support needed to fully seize energy savings opportunities, consistent internal and external messaging and financial support is required by executive management. A successful energy strategy requires accountability and transparency, but above all it requires a commitment from executive- and senior-level management to support the facility manager. Just as the company's financial performance is tracked and reported, it is important to consider the same for its energy use and carbon emissions. Increased awareness of the company's energy goals and, specifically, how each division and building contributes are key to reducing energy waste. Most companies tie employee compensation to the company's financial performance. Building owners and managers should consider tying compensation packages to the organization's energy or carbon reduction goals. An effective energy strategy requires everyone's participation. Because budgets and employee performance standards are often viewed only annually, leaders may want to reevaluate these systems to reflect longer-term views. The budget will look better today if the cheapest products are purchased; however, the future energy, maintenance, replacement, and environmental costs will be higher. Furthermore, building O&M cannot be ignored. In tough economic times, people rarely consider cutting accountants and too often consider cutting maintenance personnel. Cutting O&M budgets will increase operating costs and shorten the service life of capital equipment.

An effective supply-side and demand-side energy strategy requires the continual optimization of the acquisition of energy and the use of energy. That can easily be a full-time job, which is why hiring or appointing an energy manager or resource conservation manager who is dedicated to optimizing and measuring the company's or real estate portfolio's acquisition of and use of energy can be a wise idea. Such a move signals—internally and to clients—the importance of energy management within an organization. Energy savings opportunities seized by the energy manager more often than not pay for the cost of the position. In addition to all of the demand-side opportunities discussed in this Guide, opportunities on the supply side exist and may include reviewing rate structures, verifying utility billings, shifting peak loads to off hours when rates are lower, and analyzing and improving power factor economies.

REQUIRING A SYSTEMS MANUAL

A systems manual is the user's manual for the optimal operation of the building's systems. Without one, it only takes the loss of a single building engineer to lose all efficiencies gained.

Commercial real estate investment property is purchased by an investor based on a pro forma, or expected, financial performance. To ensure the expected performance is understood and implemented by everyone involved, annual business plans are developed. Monthly reports monitor all activities as compared with the annual business plan to ensure the property is operating as planned and stays on track with its expected financial performance. The annual business plan and pro forma financial projections both include 10-year capital budgets. Corporate facilities are governed by similar practices.

Think of a systems manual like the annual (financial) business plan, but for all of the building systems. The purpose of a systems manual is the same—to ensure the building is operating as planned and optimally. Without a plan, how does the building stay on target? What is the target? When buildings are newly constructed, drawings and operating manuals are often provided. These are often misplaced. Further, there is no record of how the building is being operated. Is it being operated optimally? Do changes in tenancy require changes in operating strategies?

The basic components of a systems manual are a prerequisite under the Leadership in Energy and Environmental Design (LEED) for Existing Buildings: Operations and Maintenance Rating System. Following are the typical components of a systems manual (PECI 2007).

- Master list of building documentation and locations
- Owner's operating requirements or building operating plan
- Sequences of operation for all control systems
- System description narrative and diagrams
- List of monitoring and control points
- List of control system alarms
- Trending capabilities
- O&M plan (includes record-keeping procedures)
- Retrocommissioning plan and/or list of energy efficiency measures (EEMs) identified by energy audit(s)
- Retrocommissioning or energy audit final report

For more detailed information about the components of a systems manual, consult Informative Annex O of *ASHRAE Guideline 0-2005, The Commissioning Process* (ASHRAE 2005b).

The systems manual contains all of the information on how the building should operate and the controls it has to assist building staff. The systems manual is the annual business plan and, therefore, is a living document that requires updating as changes occur in the building. Owners and managers should require, as a matter of policy, that the manual is routinely updated with any changes to equipment, space, use, occupancy, sequences of operation, setpoints, control strategies, schedules, etc. Systems manuals may be developed as part of the retrocommissioning process or energy audit.

FOCUSING ON EFFICIENT OPERATIONS IN O&M

Just as the monthly financial report is the tool for ensuring the company is performing as outlined in the annual business plan, the building's O&M plan should be the ongoing process of ensuring the intended performance of building systems—in accordance with the systems manual.

As discussed in previous chapters, changes in operational measures should be the first energy reduction measures taken. O&M changes incur little to no cost but can often produce significant savings. O&M opportunities are numerous and varied.

The Environmental Protection Agency's (EPA) ENERGY STAR Building Upgrade Manual is a good reference and is available for free download at www.energystar.gov/index.cfm?c=business.bus_upgrade_manual.

Some examples of changes that can be made include finding and repairing all leaks of air, steam, and water. These leaks waste surprising amounts of energy.

Also, control sensors should be calibrated periodically to ensure that readings are accurate. Control setpoints should be reviewed as well. These include space temperatures but also items such as hot- and chilled-water temperatures and economizer setpoints.

The 27-story Class-A 5 Houston Center uses an electronic preventative maintenance and tenant work order system. The maintenance team is notified via wireless device as each maintenance and inspection task becomes due. Tenants request maintenance, repair, and janitorial services online using the same system.

(Photograph courtesy of *High Performing Buildings* [Martin 2009].)

Most buildings are not open 24/7, but they use considerable energy when they are unoccupied. This presents a great opportunity to save energy because changes to thermal comfort conditions during unoccupied hours do not affect occupants. Control schedules should be reviewed to ensure that automatic shutdown and setback of building heating, ventilating and air-conditioning (HVAC) systems occurs. Morning warm-up, precooling, and optimum start controls must be properly adjusted to ensure that proper comfort conditions are achieved prior to the building opening. A midnight tour of the building is useful for spotting equipment that has been left on unnecessarily. Ventilation schedules should be modified to track occupancy. Energy can be saved by reducing excess ventilation, but care must be taken to ensure that good indoor air quality is maintained.

A change in tenant often requires remodeling of the space. If the space is repartitioned, it's key for owners and managers to review duct runs and thermostat locations and make adjustments as needed. Indiscriminate repartitioning can lead to occupant discomfort and wasted energy if the system is not modified to match the new room layout.

No-Cost Energy Efficiency Measures

The following was taken from the BOMA International BEEP course and identifies a 7% to 28% savings that can be achieved from the listed measures:

- Inspect all equipment and controls regularly to ensure that they are operating as expected to achieve energy efficiency while maintaining occupant comfort and indoor air quality. Periodically review control sequences to ensure they are consistent with current building operations and energy-efficiency strategies. Record 15-minute electric data or obtain it from the electric utility. Review electric load diagrams for unexpected anomalies or after-hours electricity use.

  Estimated savings: 2.9% to 11.5%

- Calibrate thermostats by periodically walking through the building and comparing the thermostat setting with a handheld digital thermostat. Ensure the thermostat setting equals the actual space temperature.

  Estimated savings: 0.6% to 2.9%

- Adjust dampers to bring in the least amount of outside air necessary to maintain proper air quality (within code requirements). This will minimize the need to condition outside air.

  Estimated savings: 2.9% to 5.7%

- Utilize janitorial best practices. Janitorial staff is often ignored when developing energy-saving strategies, yet they typically account for almost 25% of the weekly lighting use, which is equivalent to approximately 7% of the total building energy use. [Options to reduce the energy cost of building cleaning include:]

  - *Team Cleaning:* Janitors go through the building as a team floor by floor, and the lighting is turned on/off as they progress through the building.
  - *Coordination:* Janitors coordinate with the security crew to walk through the building and turn off equipment that was inadvertently left on by the tenants.

- *Day Cleaning*: Janitors clean during the day while the lights are already on.

Estimated savings: 0.6% to 8%

If all of these tips are implemented, energy savings of anywhere from 7% to 28% can be achieved. Therefore, it would be reasonable on average that one could expect to reduce energy use of a typical 100,000 ft^2 building, with a unit electricity cost of $0.09/kWh, by 15% and save approximately $33,000 per year. Table 5-1 shows the low and high estimates based upon changes in O&M practices. These savings can be used to finance more capital-intensive improvements such as equipment upgrades.

INVESTING IN TRAINING

The majority of, if not all, energy-saving operational measures can be implemented by building staff. Most third-party property and facility management companies have lists of operational measures that improve energy efficiency in their operating manuals. However, most are not utilized. This is because measuring, managing, and reporting energy usage has not been made a priority by those who set employee performance goals and conduct evaluations.

Further, building staff often lack information about the building (thus the need for the systems manual discussed previously) and lack necessary and building-specific training. Training is critical to efficient and successful operations. For example, owners and managers should develop an individual training plan for each member of the building team. Another option is to provide building staff an

Table 5-1. Quantifiable Results for Changes to O&M

Note: 100,000 ft^2 Blended rate = $.09/kWh Initial Energy Performance Rating = 50	**Low Estimate**		**High Estimate**	
	Energy Savings, %	**Cost Savings, $**	**Energy Savings, %**	**Cost Savings, $**
Function as Designed	2.9	$6285	11.5	$23,839
Calibrate Thermostats	0.6	$1300	2.9	$6285
Adjust Dampers	2.9	$6285	5.7	$12,353
Employ Janitorial Practices	0.6	$1322	8.0	$17,338
CUMULATIVE EFFECT	7.0	$15,192	28.1	$59,815

annual training budget of 1% to 2% of the building's annual energy cost and challenge them to reduce consumption by 10%.

Many building owners invest in energy management systems (EMSs) to manage buildings' energy usage. However, ENERGY STAR indicates there are just as many poor energy performers among buildings without EMSs as there are with EMSs (Von Nelda and Hicks 2009). This is because the majority of EMSs are not being fully utilized by building staff. Additionally, with new technologies EMSs can be very sophisticated.

At minimum, owners and managers should purchase the training module for the EMS and emphasize full utilization of the EMS to improve operational efficiency through monitoring and control strategies. Requiring maintenance and testing of the EMS should become part of the building's maintenance program. Without proper training and care, EMSs become scapegoats for comfort and control problems. Staff may eventually disable them and/or insist that new equipment is needed.

In addition to in-house training programs, the following are just a few sources for obtaining O&M energy-efficiency training:

- The BOMA International BEEP is a six-course educational series on increasing energy efficiency. Information on the program can be found at www.boma.org/TrainingAndEducation/BEEP/Pages/default.aspx.
- ASHRAE offers an abundance of training courses, guidance documents, conferences and Webinars, and a building commissioning certification program.

INSISTING ON PERFORMANCE TRACKING AND REPORTING

Without tracking data, there is no way to determine the performance of the building, the quality of the O&M practices, or the payback of capital investments made. Performance tracking is also preventative—it signals problems to building operators before energy is wasted, equipment fails, and occupants complain.

Reviewing and tracking utility bills and calculating a building's energy utilization index (EUI) provides a measure of energy performance and signals an issue but does not provide building staff with the information needed to identify the issue. An effective O&M program requires providing building staff with the diagnostic tools needed to troubleshoot and identify issues. Information is knowledge. The more information (data) that can be collected on how a building is operating (e.g., when equipment cycles on/off, space temperature, and humidity and airflow) the better it can be managed. The practice of performance tracking through data logging (i.e., collecting data through data points, trending that data, and using the software and training to understand and fully utilize the data) is often underutilized. It is not unusual for an EMS to have capabilities that never have been enabled due to a lack of understanding and training. Alternatively, a building's EMS can have insufficient data logging capabilities that need to be supplemented with low-cost additional memory or add-on modules or sensors.

As mentioned in this chapter, training in the use and optimization of the EMS cannot be overemphasized, which is why engaging a third party to commission the EMS can be a worthwhile investment.

Performance tracking and reporting to senior management signals an issue. The following indicate that potential problems are not being effectively addressed by the building's O&M practices:

- an increase in normalized energy consumption of 10% or more
- an increase in tenant complaints
- frequent equipment failures

References and Web Sites

REFERENCES

Abernathy, R., and R.E. Diaz. 2009. Creating a legacy: Skanska's Atlanta office. *High Performing Buildings* 2(1):46–55.

Anonymous. 2008. ASHRAE sets example. *High Performing Buildings* 1(4):18–26.

Architecture 2030. 2009. *The 2030 Challenge.* Santa Fe, NM: Architecture 2030. www.architecture2030.org/about.phh.

ASHRAE. 2002. *ASHRAE Guideline 14-2002, Measurement of Energy and Demand Savings.* Atlanta: American Society of Heating, Refrigerating and Air-Conditioning Engineers, Inc.

ASHRAE. 2004a. *ANSI/ASHRAE Standard 55-2004, Thermal Environmental Conditions for Human Occupancy.* Atlanta: American Society of Heating, Refrigerating and Air-Conditioning Engineers, Inc.

ASHRAE. 2004b. *Procedures for Commercial Building Energy Audits.* Atlanta: American Society of Heating, Refrigerating and Air-Conditioning Engineers, Inc.

ASHRAE. 2005a. *Advanced Energy Design Guide for Small Office Buildings.* Atlanta: American Society of Heating, Refrigerating and Air-Conditioning Engineers, Inc.

ASHRAE. 2005b. *ASHRAE Guideline 0-2005, The Commissioning Process.* Atlanta: American Society of Heating, Refrigerating and Air-Conditioning Engineers, Inc.

ASHRAE. 2006a. *Advanced Energy Design Guide for Small Retail Buildings.* Atlanta: American Society of Heating, Refrigerating and Air-Conditioning Engineers, Inc.

ASHRAE. 2006b. *ANSI/ASHRAE Standard 169-2006, Weather Data for Building Design.* Atlanta: American Society of Heating, Refrigerating and Air-Conditioning Engineers, Inc.

ASHRAE. 2007a. *2007 ASHRAE Handbook—HVAC Applications,* Chapter 35. Atlanta: American Society of Heating, Refrigerating and Air-Conditioning Engineers, Inc.

ASHRAE. 2007b. *Advanced Energy Design Guide for K–12 School Buildings.* Atlanta: American Society of Heating, Refrigerating and Air-Conditioning Engineers, Inc.

ASHRAE. 2007c. *ANSI/ASHRAE Standard 62.1-2007, Ventilation for Acceptable Indoor Air Quality.* Atlanta: American Society of Heating, Refrigerating and Air-Conditioning Engineers, Inc.

ASHRAE. 2007d. *ANSI/ASHRAE/IESNA Standard 90.1-2007, Energy Standard for Buildings Except Low-Rise Residential Buildings.* Atlanta: American Society of Heating, Refrigerating and Air-Conditioning Engineers, Inc.

ASHRAE. 2007e. *ANSI/ASHRAE Standard 105-2007, Standard Methods of Measuring, Expressing, and Comparing Building Energy Performance.* Atlanta: American Society of Heating, Refrigerating and Air-Conditioning Engineers, Inc.

ASHRAE. 2008. *Advanced Energy Design Guide for Small Warehouses and Self-Storage Buildings.* Atlanta: American Society of Heating, Refrigerating and Air-Conditioning Engineers, Inc.

ASHRAE. 2009. *Advanced Energy Design Guide for Highway Lodging.* Atlanta: American Society of Heating, Refrigerating and Air-Conditioning Engineers, Inc.

Bischak, S., and A. James. 2008. A sense of community. *High Performing Buildings* 1(2):4–19.

BOMA. 2004–2007. *Experience Exchange Report,* 2004–2007 eds. Washington, DC: Building Owners and Managers Association International.

BOMA. 2006a. BOMA Energy Efficiency Program (a series of six courses). Building Owners and Managers Association International, Washington, DC.

BOMA. 2006b. Building Energy Retrofit Program. Building Owners and Managers Association International, Washington, DC.

Briggs, R.S., R.G. Lucas, and Z.T. Taylor. 2003. Climate classification for building energy codes and standards: Part 1—Development process. *ASHRAE Transactions* 109(1):109–21.

California State Senate. AB1103. 2007–2008 Regular Session.

Council of the District of Columbia. Clean and Affordable Energy Act of 2008.

DOE. 2009a. *Building Technologies Program.* Washington, DC: U.S. Department of Energy. www.eere.energy.gov/buildings/about/htm.

DOE. 2009b. *Consumer Energy Tax Incentives.* Washington, DC: U.S. Department of Energy. www.energy.gov/taxbreaks.htm.

DOE. 2009c. *Operations and Maintenance.* Washington, DC: U.S. Department of Energy. www1.eere.energy.gov/femp/operations_maintenance.

EIA. 2003. *2003 Commercial Buildings Energy Consumption Survey.* Washington, DC: Energy Information Administration, U.S. Department of Energy.

EIA. 2006. *Annual Energy Review 2006.* Washington, DC: Energy Information Administration, U.S. Department of Energy.

EIA. 2007. *Annual Energy Review 2007.* Washington, DC: Energy Information Administration, U.S. Department of Energy.

EIA. 2008. *Annual Energy Outlook 2008 with Projections to 2030.* Washington, DC: Energy Information Administration, U.S. Department of Energy.

EIA. 2009a. *Annual Energy Outlook 2009.* Washington, DC: Energy Information Administration, U.S. Department of Energy.

EIA. 2009b. *Short-Term Energy Outlook.* Washington, DC: Energy Information Administration, U.S. Department of Energy.

Enck, H.J. 2009. Designing efficient lighting. *High Performing Buildings* 2(3):20–27.

EPA. 2008a. *2008 Professional Engineer's Guide to the ENERGY STAR Label for Commerical Buildings*. Washington, DC: U.S. Environmental Protection Agency. www.energystar.gov/ia/business/evaluate_performance/pm_pe_guide.pdf.

EPA. 2008b. *ENERGY STAR Portfolio Manager*. Washington, DC: U.S. Environmental Protection Agency. www.energystar.gov/istar/pmpam.

EPA. 2008c. *ENERGY STAR Performance Ratings: Methodology for Incorporating Source Energy Use*. Washington, DC: U.S. Environmental Protection Agency. www.energystar.gov/ia/business/evaluate_performance/site_source.pdf.

Fuller, S.K., and S.R. Petersen. 1995. *NIST Handbook 135, Life-Cycle Costing Manual for the Federal Energy Management Program*. Gaithersburg, MD: National Institute of Standards and Technology.

Hinge, A., and D. Winston. 2008. The proof is performance: How does 4 Times Square measure up? *High Performing Buildings* 1(1):30–36.

Hovorka, F. 2008. Coloring Paris lights. *High Performing Buildings* 1(3):48–55.

IEA. 2007. *World Energy Outlook 2007*, p. 4. Paris: International Energy Agency.

IES. 2000. *The IESNA Lighting Handbook: Reference and Application*. New York: Illuminating Engineering Society of North America.

IFMA. 2008. *Benchmarks V: Annual Facility Costs*. Houston, TX: International Facility Management Association.

IPCC. 2007. *Climate Change 2007: Mitigation of Climate Change,* Chapter 6. Geneva, Switzerland: Intergovernmental Panel on Climate Change.

Martin II, L. 2009. Taking care of tenants. *High Performing Buildings* 2(1): 38–44.

McKinsey & Company. 2007. *Reducing U.S. Greenhouse Gas Emissions: How Much at What Cost?* Washington, DC: McKinsey & Company.

Mills, E., H. Friedman, T. Powell, N. Boulassa, D. Claridge, T. Hassl, and M.A. Piette. 2004. The cost-effectiveness of commercial-buildings commissioning. Lawrence Berkeley National Laboratory, Berkeley, CA. http://eetd.lbl.gov/ea/emills/pubs/pdf/cx-costs-benefits.pdf.

NIST. 2009. *Building Life-Cycle Cost Program*. Gaithersburg, MD: National Institute of Standards and Technology.

North Carolina Solar Center. 2009. *Database of State Incentives for Renewables and Efficiency*. Raleigh, NC: North Carolina Solar Center. www.dsireusa.org.

NYC. 2009. *Emissions Inventory for New York City*. City of New York, New York.

Parsley, J., and A. Serra. 2009. Prague's gold standard. *High Performing Buildings* 2(3):28–39.

PECI. 2007. *A Retro-Commissioning Guide for Building Owners*, p. 80. Portland, OR: Portland Energy Conservation, Inc.

Pinches, G.E. 1984. *Essentials of Financial Management*, Chapter 12. New York: Harper & Row.

REALpac. 2007. *REALpac National Corporate Responsibility and Sustainability Guidelines*. Toronto, Ontario: Real Property Association of Canada.

Rushing, A.S., and B.C. Lippiatt. 2009. *Energy Price Indices and Discount Factors for Life-Cycle Cost Analysis*. Gaithersburg, MD: National Institute of Standards and Technology.

Schreiber, A. 2009. Better by design. *High Performing Buildings* 2(3):50–63.

UNEP FI. 2008. *Responsible Property Investing: What the Leaders Are Doing.* Geneva, Switzerland: United Nations Environment Programme Finance Initiative.

U.S. Congress. House. Energy and Independence Security Act of 2007. HR 6. 110th Cong., 1st sess.

U.S. Congress. House. Energy Improvement and Extension Act of 2008. HR 6049. 110th Con.

USCM. 2009. *U.S. Conference of Mayors Climate Protection Agreement.* Washington, DC: The United States Conference of Mayors. www.usmayors.org.

Von Nelda, B., and T. Hicks. 2009. Building performance defined: The ENERGY STAR national performance rating system. Washington, DC: U.S. Environmental Protection Agency. www.energystar.gov/ia/business/tools_resources/aesp.pdf.

West, S. 2009. Transparent performance. *High Performing Buildings* 2(2):40–49.

White, B. 2007. Energy efficiency and clean energy: Essential strategies for superior business performance. Cleaner Technology and Energy Efficiency: Structuring a Competitive, April 5, 2007, Boxborough, MA.

WRI and WBCSD. 1998. *Greenhouse Gas Protocol.* Washington, D.C., and Geneva, Switzerland: World Resources Institute and World Business Council for Sustainable Development. www.ghgprotocol.org.

WEB SITES

ASHRAE—American Society of Heating, Refrigerating
and Air-Conditioning Engineers, Inc.
    www.ashrae.org

Building Owners and Managers Association International
    www.boma.org

DOE—U.S. Department of Energy, Building Technologies Program
    www.eere.energy.gov/buildings/about.html

DOE—U.S. Department of Energy, Tax Breaks
    www.energy.gov/taxbreaks.htm

ENERGY STAR for government
    www.energystar.gov/government

ENERGY STAR Portfolio Manager
    www.energystar.gov/istar/pmpam

Energy Information Administration
    www.eia.doe.gov

California Energy Commission
    www.energy.ca.gov

Federal Energy Management Program
    www1.eere.energy.gov/femp

North Carolina Solar Center
    www.dsireusa.org

Office of the Federal Environmental Executive
    http://ofee.gov